Maria Elizete Kunkel

Biomechanical models of the human thoracic and lumbar spine

Maria Elizete Kunkel

Biomechanical models of the human thoracic and lumbar spine

A statistical approach for prediction of anatomical parameters from radiographic images

Südwestdeutscher Verlag für Hochschulschriften

Impressum/Imprint (nur für Deutschland/only for Germany)
Bibliografische Information der Deutschen Nationalbibliothek: Die Deutsche Nationalbibliothek verzeichnet diese Publikation in der Deutschen Nationalbibliografie; detaillierte bibliografische Daten sind im Internet über http://dnb.d-nb.de abrufbar.
Alle in diesem Buch genannten Marken und Produktnamen unterliegen warenzeichen-, marken- oder patentrechtlichem Schutz bzw. sind Warenzeichen oder eingetragene Warenzeichen der jeweiligen Inhaber. Die Wiedergabe von Marken, Produktnamen, Gebrauchsnamen, Handelsnamen, Warenbezeichnungen u.s.w. in diesem Werk berechtigt auch ohne besondere Kennzeichnung nicht zu der Annahme, dass solche Namen im Sinne der Warenzeichen- und Markenschutzgesetzgebung als frei zu betrachten wären und daher von jedermann benutzt werden dürften.

Coverbild: www.ingimage.com

Verlag: Südwestdeutscher Verlag für Hochschulschriften GmbH & Co. KG
Heinrich-Böcking-Str. 6-8, 66121 Saarbrücken, Deutschland
Telefon +49 681 37 20 271-1, Telefax +49 681 37 20 271-0
Email: info@svh-verlag.de

Approved by: Ulm, Ulm University, Diss., 2011

Herstellung in Deutschland (siehe letzte Seite)
ISBN: 978-3-8381-3172-6

Imprint (only for USA, GB)
Bibliographic information published by the Deutsche Nationalbibliothek: The Deutsche Nationalbibliothek lists this publication in the Deutsche Nationalbibliografie; detailed bibliographic data are available in the Internet at http://dnb.d-nb.de.
Any brand names and product names mentioned in this book are subject to trademark, brand or patent protection and are trademarks or registered trademarks of their respective holders. The use of brand names, product names, common names, trade names, product descriptions etc. even without a particular marking in this works is in no way to be construed to mean that such names may be regarded as unrestricted in respect of trademark and brand protection legislation and could thus be used by anyone.

Cover image: www.ingimage.com

Publisher: Südwestdeutscher Verlag für Hochschulschriften GmbH & Co. KG
Heinrich-Böcking-Str. 6-8, 66121 Saarbrücken, Germany
Phone +49 681 37 20 271-1, Fax +49 681 37 20 271-0
Email: info@svh-verlag.de

Printed in the U.S.A.
Printed in the U.K. by (see last page)
ISBN: 978-3-8381-3172-6

Copyright © 2012 by the author and Südwestdeutscher Verlag für Hochschulschriften GmbH & Co. KG and licensors
All rights reserved. Saarbrücken 2012

A statistical approach to predict subject-specific morphometry of the human thoracic and lumbar spine from radiographic images

Dr. Maria Elizete Kunkel

Contents

1. Introduction   1

2. Prediction equations for human thoracic and lumbar vertebral morphometry   8

3. Prediction of the human thoracic and lumbar articular facet joint morphometry
   from radiographic images   13

4. Morphometric analysis of the relationships between intervertebral disc and
   vertebral body heights: An anatomical and radiographic study of the human thoracic
   spine   18

5. Conclusion   23

6. Summary   25

7. References   26

8. Papers   36

Nomenclature

2D	two-dimensional
3D	three-dimensional
ADH	anterior disc height
AFJ	articular facet joints
AP	anterior-posterior
CT	computed tomography
CV	coefficient of variation
EPDI	end-plate depth inferior
EPDS	end-plate depth superior
EPII	end-plate inclination inferior
EPIS	end-plate inclination superior
EPWI	end-plate width inferior
EPWS	end-plate width superior
FE	finite element
FHIL	facet height inferior left
FHIR	facet height inferior right
FHSL	facet height superior left
FHSR	facet height superior right
FWIL	facet height inferior left
FWIR	facet height inferior right
FWSL	facet width superior left
FWSR	facet height superior right
IFHL	interfacet height left
IFHR	interfacet height right
IFWI	interfacet width inferior
IFWS	interfacet width superior
L1	1st lumbar vertebra
L2	2^{sd} lumbar vertebra
L3	3th lumbar vertebra
L4	4th lumbar vertebra
L5	5th lumbar vertebra
MDH	middle disc height

mm	millimetres
MRI	magnetic resonance imaging
P	statistical significance level
PDH	posterior disc height
PHL	pedicle height left
PHR	pedicle height right
PSIL	pedicle sagittal inclination left
PSIR	pedicle sagittal inclination right
PTIL	pedicle transverse inclination left
PTIR	pedicle transverse inclination right
PWL	pedicle width left
PWR	sagittal width right
R^2	correlation coefficient
SAIL	sagittal angle inferior left
SAIR	sagittal angle inferior right
SASL	sagittal angle superior left
SASR	sagittal angle superior right
SCD	sagittal canal depth
SCW	spinal canal width
SD	standard deviation
SE	standard error
SPL	spinous process length
T1	1st thoracal vertebra
T2	2sd thoracal vertebra
T3	3th thoracal vertebra
T4	4th thoracal vertebra
T5	5th thoracal vertebra
T6	6th thoracal vertebra
T7	7th thoracal vertebra
T8	8th thoracal vertebra
T9	9th thoracal vertebra
T10	10th thoracal vertebra
T11	11th thoracal vertebra
T12	12th thoracal vertebra

TAIL	transverse angle inferior left
TAIR	transverse angle superior right
TASL	transverse angle superior left
TASR	transverse angle inferior right
TIVD	thoracic intervertebral disc
TPD	transverse process depth
TPW	transverse process width
VBD	vertebral body depth
VBHA	vertebral body height anterior
VBHP	vertebral body height posterior
VBW	vertebral body width

1. Introduction

During the last decade there have been considerable developments in new techniques of surgical treatments to stabilize and correct the human spine. Many approaches have been proposed for patient-specific modeling of the human spine to explore the correction of spinal deformities, such as scoliosis, by spinal instrumentation. The current increased interest in biomechanical models of the spine and spinal implants calls for a detailed knowledge of spine morphometry and relationships between geometrical dimensions of the vertebrae and the intervertebral discs. The malfunction of these structures due to spinal pathologies or accidents represents worldwide a high-cost for medical care (Hansson & Hansson 2000, Wenig et al. 2009). Accurate anatomical descriptions of the size, shape and orientation of the main structures of the human vertebrae and intervertebral discs are necessary for a variety of approaches and objectives:

(1) The identification of clinical situations that are related to the morphometry of the spine structures, such as the incidence of low-back pain related to the spinal canal size (Porter et al 1980); the incidence of disc herniation dependent on the shape of the lumbar vertebrae (Frederick et al. 2001); rotational coupling in the spine or disc failure due to the orientation of the articular facets (Farfan & Sullivan 1967; Adams & Hutton 1983; Duncan & Ahmed 1989; Abumi et al. 1990). (2) The development of anthropological and forensic approaches for the identification of human remains where precise information about quantitative aspects of vertebral and intervertebral disc morphometry is required (Jason & Taylor 1995; Kósa & Castellana 2005; Yu & Lee 2008). (3) The understanding of both the normal and abnormal morphology of the spine in cases of spine disorders such as scoliosis and kyphosis (Manns et al 1996; Parent et al. 2004). (4) The development and use of implantable devices for spinal instrumentation, e.g. design of transpedicular fixation devices based on the size and orientation of the pedicles (Krag et al 1986, 1988; Berry et al. 1987; Zindrick et al 1987; Scoles et al. 1988; Abuzayed et al. 2010); analysis of the vertebral morphometry in idiopathic scoliosis treated by pedicle screw instrumentation (Lilijenqvist et al. 2000; Parent et al. 2004); development of artificial intervertebral discs (Aharinejad et al. 1990). (5) To compare morphometric similarities and differences of spinal structures of animals which are used as experimental models relative to the humane spine, e.g. vertebrae (Wilke et al. 1996, 1997a, 1997b; McLain & Yerby, 2002) and intervertebral discs (O'Connell et al. 2007). (6) To build a base of anatomical data for the construction of accurate parameterized mathematical models of the human spine (Nissan & Gilad, 1984).

Several quantitative studies have investigated the external geometry of the vertebrae and intervertebral discs of different regions of the human spine. Morphometric measurements with cadaveric vertebrae have been taken directly from bony specimens or have been obtained from medical images (e.g. plain radiographs, computed tomography (CT) or magnetic resonance imaging (MRI)). However, these *in vitro* studies have focused on only a specific anatomical structure such as the vertebral body (Larsen 1985a, 1985b; Hall et al. 1998), spinal canal (Huizinga et al. 1952; Jones 1978; Eisenstein 1976, 1977, 1983; Postacchini et al. 1983), pedicle (Zindrick et al. 1987; Krag et al. 1988; Marchesi et al. 1988; Moran et al. 1989; Olsewski et al. 1990) and articular facet joints (van Schaik et al. 1985a; Cotterill et al. 1986; Berry et al. 1987; Scoles et al. 1988; Ahmed et al. 1990; Boszczyk 1997; Ebraheim et al. 1997; Laporte et al. 2000; Masharawi et al. 2004, 2005, 2007a, 2007b; Wang & Yang 2009), in a limited set of structures (Berry et al. 1987), in a limited segment of the spine such as thoracic (Cotterill et al. 1986; Scole et al. 1988; Aharinejad et al. 1990) or lumbar (Berry et al. 1987; Boszczyk 1997; Semaan et al. 2001), or in a specific population group such as South African negroes (Eisenstein, 1977), Italians (Postacchini et al. 1983), Japanese (Nojiri et al. 1990), Koreans (Lee et al. 1995) and Indians (Singh et al. 2011). The most complete collection of quantitative three-dimensional (3D) surface anatomy of the main vertebral parameters of the entire human spine was provided in Panjabi et al. (1991; 1992; 1993) (Figure 1). Since this dataset has been used in the current research, a detailed description of the measured parameters will be provided in the Chapters 2 and 3.

In vivo, accurate assessment of the geometry of the vertebrae has been typically obtained through segmentation and 3D reconstruction of CT data (e.g. van Schaik et al. 1985b) or MRI (e.g. Dai 2001). These techniques provide accurate geometrical assessment, but since this process cannot be totally automated, a long processing time and considerable computational power is required for the manual or semi-automatic segmentation of the images. Moreover, as well as besides being an expensive method, for CT imaging the subject has to be submitted to relatively high doses of ionizing radiation, and since it is performed with the patient in a supine position, changing in the spine posture should be considered (Yazici et al. 2001).

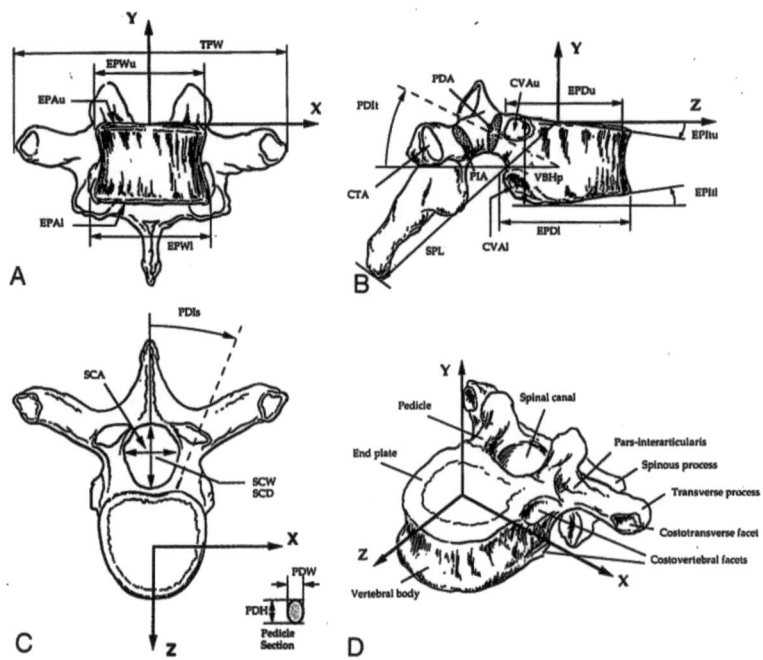

Figure 1: Schematic representation of anatomical dimensions of the human thoracic vertebra. Tree orthogonal views, front (A), side (B), top (C), and an isometric view showing the coordinate system used to define these dimensions (D). (Panjabi et al. 1991).

Geometrical dimensions of the human lumbar intervertebral discs are found in several studies (e.g. Farfan et al. 1972; Nissan & Gilad, 1986; Amonno-Kuofi 1991; Eijkelkamp 2002; Shao et al. 2002; van der Houwen et al. 2010). However, the morphometry of the thoracic discs has received limited attention despite the thoracic spine being the most common site for spinal deformities such as kyphosis, lordosis and scoliosis. For example, accurate anatomical data on the disc heights including all levels of the thoracic spine of a representative adult population are very scarce. Few *ex vivo* studies have been carried out on the thoracic disc due to the difficulty in obtaining intact human specimens. Previous studies showed limitations either in accuracy, study population, parameters recorded, or disc level. Todd & Pyle (1928) measured only anterior heights of discs of male cadavers. Pooni et al. (1986) used only a few elderly cadavers, and the data were published only as a percentage of the total spine height. Radiographic measurements by Goh et al. (1999) and Giles & Singer (2000) were performed for the investigation of thoracic kyphosis, but the anterior and posterior heights of the disc were not provided and only a segmental trend was reported. The measurements of thoracic disc heights by Pooni et al. (1986), Goh et al. (1999) and Giles & Singer (2000) were performed on plain radiographs. The repeatability of these measurements have been questioned due to a lack of the requisites needed to perform geometric measurements with relative accuracy such as the use of a standard vertebral position, control of the film-specimen-focus distances and optimal visualization of the bony landmarks (Pope et al. 1977; Andersson et al. 1981). Furthermore, in some of these investigations, errors due to radiographic magnification bias or inter- and intraobserver reliability of the radiographic measurements were not taken into account (Hurxthal 1968; Manns et al. 1986; Pooni et al. 1986). For example, Hurxthal (1968) and Manns et al. (1986) measured anterior disc height using radiographs of female patients but only a limited number of thoracic levels (from T5-6 to T11-12) were investigated.

Planar radiography is the technique frequently used in clinical diagnosis and for evaluation of spinal deformities (Dupuis et al. 1985; Carman et al. 1990). Due to this fact, some approaches were proposed in recent decades to extract geometrical information of the spine from radiographs (Gilad & Nissan, 1986). The main advantage of these approaches is that radiography enables capture of the entire bony structures of the spine while being much less invasive and expensive than CT. The disadvantage is that radiographs are two-dimensional (2D) projections and do not allow direct assessment of 3D information of the spine's structures. Therefore, the main problem that arises is in determining the real dimensions of the structures captured by X-rays as well as to identify their shape, position and orientation in all planes. The stereo-radiographic approach is an alternative procedure that has been proposed for 3D reconstruction of the spinal structures (e.g. Dansereau & Stokes, 1988). In this case, one film plane is used, and two exposures are made with different positions of the X-ray source (Aubin et al. 1997; Petit et al. 1998; Cheriet et al. 1999a, 1999b; Dumas et al. 2003, 2004; Kadoury et al. 2007a, 2007b). It provides good parameterized information about the vertebrae, but it is also time-consuming (due to the long process of identification of numerous anatomical landmarks) and resource-consuming requiring specific software and hardware. The most accurate methods to provide vertebral parameters using radiographs are still the semi-automatic approaches (Pomero et al. 2004; Dumas et al. 2008). Benameur et al. (2005) proposed an automatic approach but it was limited to the lower lumbar spine.

Statistical approaches for prediction of the human spine morphometry

Several alternative approaches have been used to provide accurate anatomical data related to the morphometry of the human spine. Previous studies have investigated whether vertebral relationships could be used to predict vertebral morphometry using the statistical correlations between anatomical dimensions of the human vertebral structures. A statistical approach based on these relationships could eliminate the need for preliminary processing of medical images such as CT to provide anatomical data on the human spine. However, statistical analyses performed using simple linear regressions between the main parameters of the vertebrae and intervertebral discs (e.g. Gilad & Nissan (1986); Scoles et al. (1988); Skalli et al. (1993); Maurel et al. (1997); Breglia (2006)) found low or no correlations for some important parameters, such as the dimensions of pedicles. Scoles et al. (1988) for example, described the posterior structures of the vertebrae as being highly variable and largely unpredictable. Some studies used multiple linear regression analyses (e.g. Lavaste et al. 1992; Laport et al. 2000) to provide methods for the reconstruction of the human vertebrae from two radiographs (anterior-posterior (AP) and lateral). They used statistical relationships between vertebral dimensions to generate vertebral data for parameterized models of the spine. However, the models based on the generated parameters have a very simplified geometry and the ability to predict vertebral parameters with this approach has been questioned because no validation using a second set of experimental measurements was provided (Lavaste et al. 1992; Laport et al. 2000).

In all the studies described above, linear regressions were used to find the correlation between the vertebral or intervertebral disc parameters. No study was conducted to investigate whether a non-linear regression (e.g. exponential, logarithmic or polynomial) could provide better results. Furthermore, the relationships between anatomical dimensions of the vertebrae and intervertebral discs including the thoracic and lumbar spine have never been investigated. To the author's knowledge, to date no report has presented a statistical approach for useful predictors of intervertebral disc dimensions based on the size of the vertebral bodies. Such quantitative analysis could provide anatomical data for patient-specific modelling of the spine from only one or a few initial vertebral dimensions.

Numerous studies have reported the impact of the use of the finite element (FE) method in conjunction with experimental studies to investigate the mechanical behaviour of the normal and pathological human spine (e.g. Gilbertson et al. 1995; Goel & Gilbertson, 1995). In FE modelling, an anatomical structure is divided into a finite number of elements that can interact with one another through their points of attachment. The important geometrical and material properties of this structure may be incorporated into the FE model and different types of structural analyses (e.g. static, dynamic) can be carried out by simulating a variety of clinical situations.

During the last decade, many biomechanical models have been proposed in order to simulate surgical correction of scoliosis supported by spinal instrumentation (Lafage et al. (2002, 2004); Duke et al. (2004); Dumas et al. (2005)). Scoliosis is a complex 3D structural deformity of the spine that involves morphological deformation of the vertebrae (e.g. asymmetrical pedicle length, spinous process deviation, facet joint asymmetry and intervertebral discs (Stokes, 1994)). Such biomechanical models have been considered as a way to overcome the limitation of the clinical investigations and laboratory experiments with cadavers and animals. Since the accuracy and reliability of the mathematical modelling of the human spine depends directly on the geometry of the spinal structures, reliable biomechanical simulations of the behaviour of the human spine require 3D modelling of the complex anatomy of the vertebrae and intervertebral discs (Robin, 1994).

Consider the articular facet joints, which are vertebral structures that play an important role in the biomechanics of the spine, because they transmit a significant percentage of the spinal loading, and provide translational, rotational and axial spinal stability (Lorenz et al. 1983; Adams & Hutton, 1993; Onan et al. 1998). The physiological motion of the spine is thus strongly dependent on the shape, position and orientation of the AFJ (Taylor & Twomey, 1986; White & Panjabi, 1990). This issue has been investigated in normal and pathological conditions through FE models of the spine created from 3D reconstruction of CT images (Shirazi-Adl 1991, 1994; Zander et al. 2003; Schmidt et al. 2008a, 2008b, 2009). However, in cases of simulations of spinal deformities such as scoliosis, morphological changes in the anatomy of the AFJ such as facet height and width, and facet angles should be implemented in the computer models (Millner & Dickson, 1996; Maurel et al. 1997; Parent et al. 2002; Aebi, 2005). Since it is difficult to appropriately to modify the geometry obtained from CT reconstructions, FE models with a simple but parameterized geometry could be an alternative to these complex models. In parameterized models of the spine, different vertebral and disc geometries may be created merely by changing the input data or parameters that represent the anatomical dimensions (Maurel et al. 1997).

Using an adequate statistical approach, the initial measurements of only a few dimensions on radiographs of patients (e.g. vertebral body height) could be enough for a rapid prediction of other dimensions (e.g. articular facet size or orientation) by using the relationships between them. Most surgical procedures for spinal deformity corrections are based on the average values of vertebral dimensions of a healthy spine, without taking into account the fact that each patient has a specific morphometry. A statistical approach allowing rapid acquisition of geometrical parameters specific to a given subject could be used both for clinical evaluation and for parameterized subject-specific modelling of the spine for biomechanical research.

The objective of this cumulative PhD-thesis was to develop a statistical approach for the acquisition of subject-specific morphometry of the main thoracic and lumbar spinal dimensions from radiographic images. Some research questions related to spine morphometry that have not been sufficiently addressed in the previous studies have been formulated as hypotheses:

(i.) The main vertebral and intervertebral disc dimensions can be measured on lateral radiographs with ease and accuracy.
(ii.) There are unique vertebral dimensions that show good correlation between all other vertebral and intervertebral disc dimensions.
(iii.) The result of these correlations depends on the inclusion or exclusion of given spinal levels (e.g. the transition regions) or the consideration of only a particular region of the spine (e.g. thoracic or lumbar).
(iv.) A nonlinear regression would be able to better describe these correlations than a linear regression by means of an equation.
(v.) It is possible to predict accurately vertebral and disc morphometry of the thoracic and lumbar spine from only one initial dimension
(vi.) It is possible to predict linear and angular dimensions of the vertebrae and intervertebral discs that are not visible on lateral radiographs.
(vii.) It is possible to make specific predictions of the morphometry of the vertebrae and intervertebral discs for an individual using these equations.
(viii.) It is possible to validate these prediction equations.

In order to test these hypotheses, three studies concerned with the morphometry of the vertebrae and intervertebral discs of the thoracic and lumbar spine were carried out using a linear and nonlinear statistical approach. In Chapter 2, a morphometric analysis performed with experimental vertebral data is described. Prediction equations were generated to estimate the main anatomical dimensions of the thoracic and lumbar vertebrae from the unique radiographic measurement of heights the vertebral bodies. A similar approach is presented in Chapter 3 to predict the articular facet joints morphometry taking into consideration the anatomical variation between the thoracic and lumbar vertebrae. Chapter 4 describes an *ex vivo* study for direct and radiographic measurements of the human thoracic intervertebral disc height. The radiological measurement error was calculated. Equations were generated for the prediction of the intervertebral disc height from measurement of the heights of vertebral bodies using a similar statistical approach.

2. Prediction equations for human thoracic and lumbar vertebral morphometry

Published in: Kunkel ME, Schmidt H, Wilke H-J (2010) Prediction equations for human thoracic and lumbar vertebral morphometry. *J Anat* **216**, 320-328.

Statistical correlations between anatomical dimensions of vertebral structures have indicated a potential that could be applied to provide geometric data for the development of simplified geometrical models of the spine while excluding the need for preliminary processing of medical images (e.g. Scoles et al. 1988; Lavaste et al. 1992; Laporte et al. 2000; Breglia, 2006). In this study, linear and nonlinear regressions were performed for the generation of prediction equations for 20 vertebral parameters of the human thoracic and lumbar spine as a function of only one given vertebral parameter that could be measured from radiographs.

The complex geometry of the thoracic and lumbar vertebrae from the level T1 to L4 was simplified as 21 vertebral parameters related to the end-plate, vertebral body, pedicle, spinal canal, and spinous and transverse processes (Figure 2). The vertebral data were obtained from the cadaveric studies of Panjabi et al. (1991; 1992; 1993). This data set represents an average of a non pathological adult population including the means values of 15 linear and 6 angular parameters of 12 human cadavers. The mean age of the subjects was 46.3 years (range: 19 – 59 years), weight was 67.8 kg (range: 54 – 85 kg), height was 167.8 cm (range: 157 – 178 cm), and the male / female ratio was 8:4.

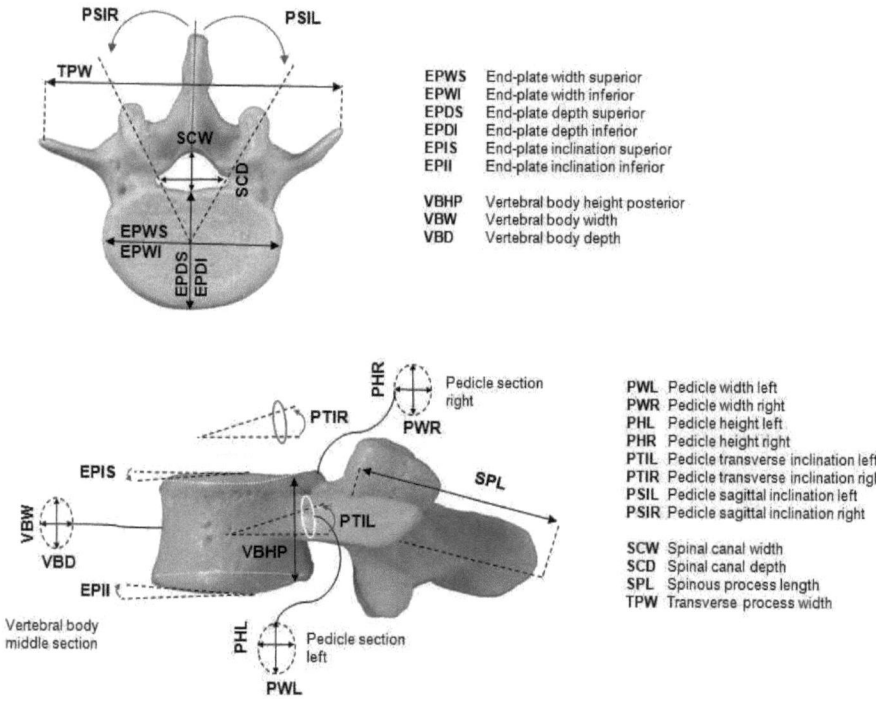

Figure 2: Schematic representation of the 21 vertebral parameters that were considered for linear and nonlinear regression analyses.

Each of these 21 vertebral parameter was considered and tested individually as a predictor variable. The parameters were individually regressed against the possible predictor variable by a least-squares estimation process. Based on the level of correlation with the other parameters and ease of measurement on lateral radiographs, the parameter VBHP was chosen and the statistical analyses described in this study are related to this parameter. Linear and nonlinear regression analyses were employed to find the best functions to fit each parameter in a prediction equation. The following hypotheses were tested: (i) a function could not fit the data significantly better than a horizontal line; (ii) a second-order equation could not fit the data significantly better than a linear equation; (iii) a third-order equation could not fit the data significantly better than a second-order equation, and so on.

The statistical procedure performed on each parameter corresponds to four-steps: Consider the parameters EPWS and VBHP (Figure 3): (1) Least-squares estimation was used to find equations to describe the relationship between EPWS and VBHP. Initially, a linear regression was performed, fitting an equation of the form $y = C_1 + C_2 x$ to the data. The variance of the EPWS was determined by the R^2 value. A logarithmic and an exponential curve with equations of the form $y = C_1 + C_2 \ln(x)$ and $y = C_1 e^{C_2 x}$, respectively, were then used to test the increase in R^2. Next, polynomial equations including more coefficients (C_1, C_2, C_3, C_4, etc.) were used to find the best fit. This was continued until adding another higher-order term did not significantly increase R^2. (2) An analysis of variance was performed to select the best prediction equation. High values of R^2 associated with a P-value < 0.01 indicated the third-order polynomial as the best-fitting equation. (3) It was evaluated how the selected nonlinear equation fits the EPWS data significantly better than a linear equation by superimposing experimental values. (4) The predictability of the best-fitting equation was evaluated using experimental data from two further datasets (Berry et al. (1987); Scoles et al. (1988)).

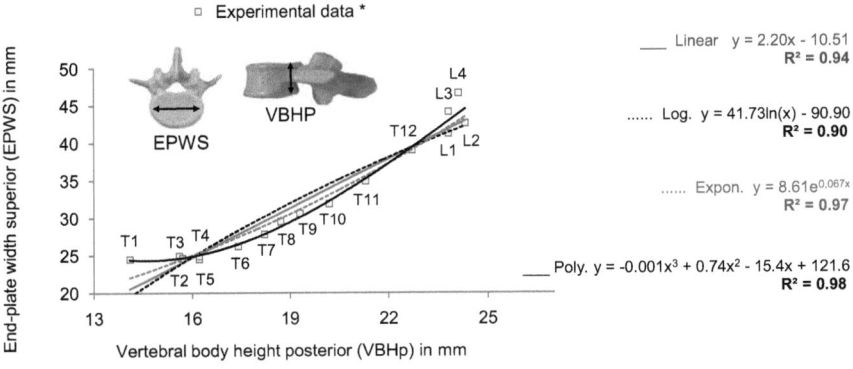

Figure 3: Description of the statistical procedure for the parameters EPWS and VBHP of the thoracic and lumbar spine. Set of prediction equations generated from linear, logarithmic, exponential and polynomial regression analyses (y is the value of EPWS and x is the VBHP on each vertebral level).

The linear, exponential and logarithmic regressions provided significant predictions of parameters related to the anterior vertebral structures from the values of VBHP. However, third-order polynomial prediction equations allowed an improvement on these predictions (P-values < 0.001), e.g., end-plates and spinal canal (R^2; 0.970-0.995) as well as pedicle heights and the spinous process (R^2; 0.811-0.882), in addition to a reasonable prediction of the parameters of the posterior vertebral structures which have shown a low or no correlation with VBHP in previous studies, e.g., pedicles inclination and transverse process (R^2; 0.514-0.693) (ANOVA). The inclusion of more than four coefficients increased the R^2 values but the obtained correlations did not significantly improve the parameter predictions. The polynomial predictions were generally within or close to the regions of the 95% confidence intervals of the experimental data of Panjabi et al. (1991; 1992) (Figure 4). Comparisons of the theoretical predictions with two sets of experimental data indicated that the predictions generally agree well with these data (Figure 5).

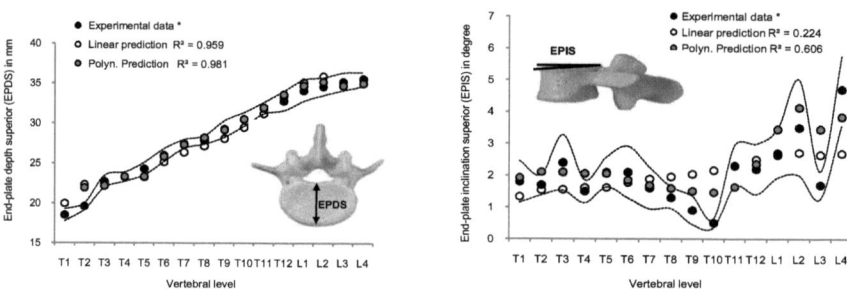

Figure 4: Linear and polynomial predictions of two selected vertebral parameters (EPDS and EPIS) superimposed on experimental data of Panjabi et al. (1991a, 1992)*. Dotted curve indicates standard deviation of the experimental data.

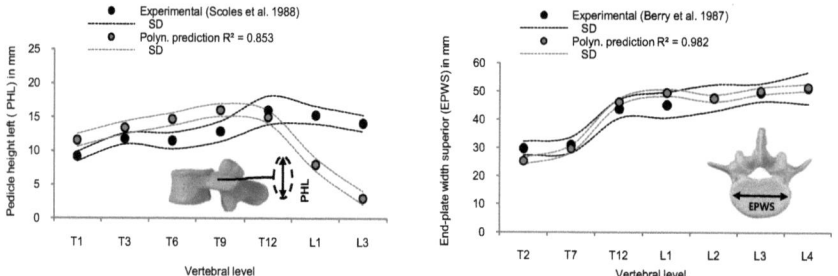

Figure 5: Comparison of two predicted vertebral parameters (PHL and EPWS) with corres-ponding experimental data in selected vertebrae. The means and 95% confidence intervals (dotted lines) of the experimental and predicted values are shown.

The present study provided a time efficient approach for the prediction of mor-phometry of the human thoracic and lumbar vertebrae. It allows a better understanding of statistical correlations between these vertebral dimensions and could be used to provide data for geometrical modeling of the human vertebrae. It requires the measurement of only one parameter per vertebra (VBHP) from a lateral radiograph, and the set of developed prediction equations. Vertebral models based on this type of parameterized geometry could be used in biomechanical studies which require variation of the geometry, such as in spinal deformations, including scoliosis. All correlation coefficients found in the current study were considerably better than the values obtained in previous studies. Scoles et al. (1988) reported the impossibility to establish useful predictors for pedicle dimensions based on the size of the vertebral body; Lavaste et al. (1992) developed a method to reconstruct lumbar vertebral geometry from radiographs using multiple-linear regression analysis, requiring six given parameters per vertebra to predict the vertebral geometry with a digitalization process that showed a relative error of approximately 15%; Laporte et al. (2000) required the measurement of 15 parameters per thoracic vertebra from radiographs in order to predict each other parameter; Breglia (2006) used only linear regressions with the same data from Panjab as used in the current study and could not predict the parameters related to posterior structures.

The differential in the current study in relation to these previous studies is based on two facts that were considered in the statistical analyses: (1) The data corresponding to the vertebral level L5 were not included in the regressions. Several studies reporting the remarkable morphological differences found in the structures of this vertebral level when compared with the neighbouring vertebrae as due to L5 to being in a transition zone, from lumbar to sacral region (Berry et al. 1987; Zindrick et al. 1987; Scoles et al. 1988; Panjabi et al. 1989; 1992). (2) Linear and nonlinear regressions were tested and compared. The relationships between the vertebral variables follow a curved line, not a straight line. Fitting a nonlinear equation such as polynomial regression provides better results because polynomial equations can be used to create a generic curve through the data points; more coefficients better fit the data.

There are three advantages to the use of the developed approach to predict 20 vertebral parameters per vertebral level (T1-L4) with radiographic measurement of the VBHP. This reduces by approximately 94% the need for individual measurement of 20 parameters per vertebral level; it provides information about the angular parameters that were not measurable in radiographs; and since this statistical approach is based on the geometrical relationships between the vertebrae, the measurement of the VBHP of a subject provides parameters taking into account geometric characteristics specific of the subject.

3. Prediction of the human thoracic and lumbar articular facet joint morphometry from radiographic images

Published in: Kunkel ME, Schmidt H, Wilke H-J (2011) Prediction of the human thoracic and lumbar articular facet joint morphometry from radiographic images. *J Anat* **218**, 191-201.

Although it is well known that some dimensions of the articular facet joints (AFJ) (e.g. facet height / width or facet angles) play a major role in spinal deformities such as scoliosis, little is known about statistical correlations between these dimensions and the size of the vertebral bodies. Breglia (2006) investigated only two AFJ parameters and a poor correlation between them and the VBHP was found. Other studies that investigated statistical correlations between AFJ parameters and many other parameters reported that a unique parameter could not provide an accurate prediction (Lavaste et al. 1992; Skalli et al. 1993; Maurel et al. 1997; Laport et al. 2000). It would be of clinical interest to use such relationships for subject-specific predictions of AFJ parameters for mathematical modelling of the spine from a single dimensions measurable from radiographs. The aim of this study was to generate prediction equations for 20 parameters of the human thoracic and lumbar AFJ from T1 to L4 as a function of only one given parameter, the VBHP.

In order to carry out statistical analyses, the anatomical measurement of the VBHP and 20 anatomical measurements related to the size and orientation of the human thoracic (T1-12) and lumbar (L1-4) AFJ were selected from the data of Panjabi et al. (1991; 1992; 1993) (Figure 6). To perform linear and nonlinear regression with these data, the same methodology used in the Chapter 2 (Kunkel et al. (2010)) was adopted with the parameter VBHP as the predictor variable. However, due to the existence of a considerable difference between the morphometry of the AFJ in the thoracic and lumbar regions, a modification was introduced in order to achieve a better prediction of the AFJ parameters. It refers to separate regressions using thoracic (T1-12) and lumbar (L1-4) data together, thoracic data alone, and lumbar data alone. (1) The identification of the best prediction equations was based on values of $R^2 > 0.5$ associated with a probability level of P-value < 0.05 and a standard error of the estimate (SE) < 30% of the standard deviations of the experimental data of Panjabi et al (1993). For validation, the theoretical predictions were plotted against the experimental data of Panjabi et al. (1993) and Cotterill et al. (1986).

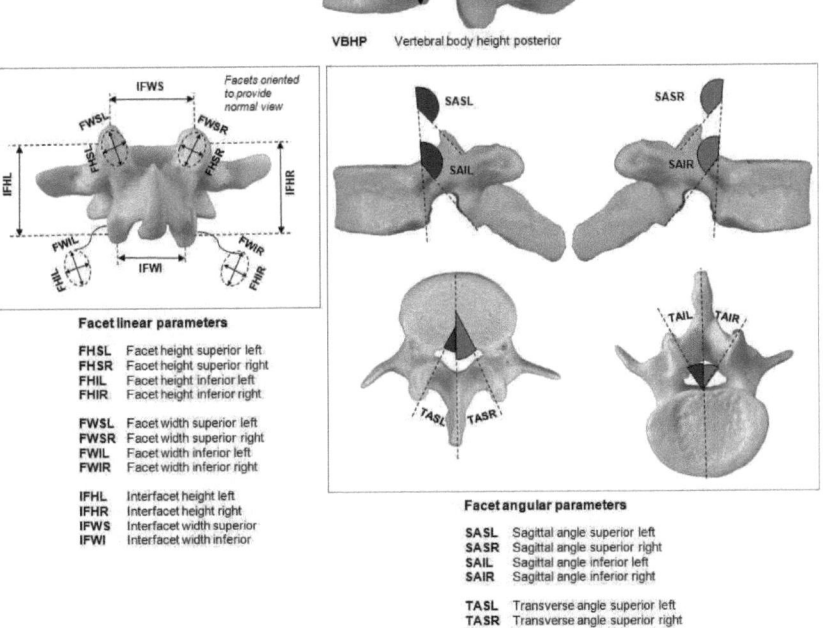

Figure 6: Schematic representation of the anatomical parameters that were considered for linear and nonlinear regression analyses: Vertebral body height posterior and dimensions of the articular facet joints linear and angular.

Third-order polynomial regressions in contrast to the linear, exponential, logarithmic and polynomial regressions with other orders provided the best results with significant correlations between each of the AFJ parameters and VBHP. The polynomial regressions using the thoracic and lumbar data together showed variable correlations with VBHP (R^2, from 0.516 to 0.950) providing significant prediction equations for all selected AFJ parameters from VBHP (an exception was FWSL). When considering the polynomial regressions for only the thoracic data (T1-12), this resulted in improvements for predictions of 70% of the thoracic AFJ parameters such as an increase of R^2 (from 0.650 to 0.973), an increase in the significance and a decrease of SE. Polynomial regressions considering only lumbar data (L1-4) did not reach the minimum criteria required for the selection of the prediction equations and the polynomial coefficients obtained from the regressions using the thoracic and lumbar data together were used. Third-order polynomial regressions provided moderate to high correlations between the AFJ parameters and VBHP (e.g. facet height superior and interfacet width (R^2, 0.605-0.880); facet height inferior, interfacet height and sagittal / transverse angle superior (R^2, 0.875 - 0.973)). Different correlations were found for facet width and transverse angle inferior in the thoracic (R^2, 0.703 - 0.930) and lumbar (R^2, 0.457 - 0.892) regions.

Comparisons with experimental data of Panjabi et al. (1993) showed that the best predictions were found for FHIR with a mean percent error of approximately 14% in T12 and a maximal error of 7% in all other levels. The largest error of approximately -17% was found for FHSL predictions in the lumbar level (Figure 7). High correlations were found for TASL with VBHP for all levels (R^2, 0.785–0.938). TAIR displayed a poor correlation (R^2 = 0.526) for the lumbar levels (Figure 7). The predictions with the data of Panjabi et al. (1993) showed the largest mean percent error of approximately 10%. Comparison of the predictions of the interfacet heights and width with experimental data of Cotterill et al. (1986) indicated mean percent errors < 16%, with the exception of the thoracolumbar junction (T12 - L1).

The best predictions of facet orientations were found for the transverse angles, probably because these parameters show very little variability within the vertebrae from T1 to L4. Notably in the midthoracic region (i.e. T3-8) excellent predictions with errors < 10% could be achieved for most parameters of the AFJ. In contrast, in the thoracolumbar junction (T12-L1) were found predictions with errors of up to -15% for all sagittal angles. This was due to the large variability of this region within individuals being the AFJ either frontally oriented as in the thoracic vertebrae or sagittally oriented as in the lumbar vertebrae. This is in accordance with Goel & Weinstein (1990) and Masharavi et al. (2004) who showed that the morphology of the first lumbar vertebra is distinct from the other vertebrae with a transition from the typically thoracic to the lumbar vertebra.

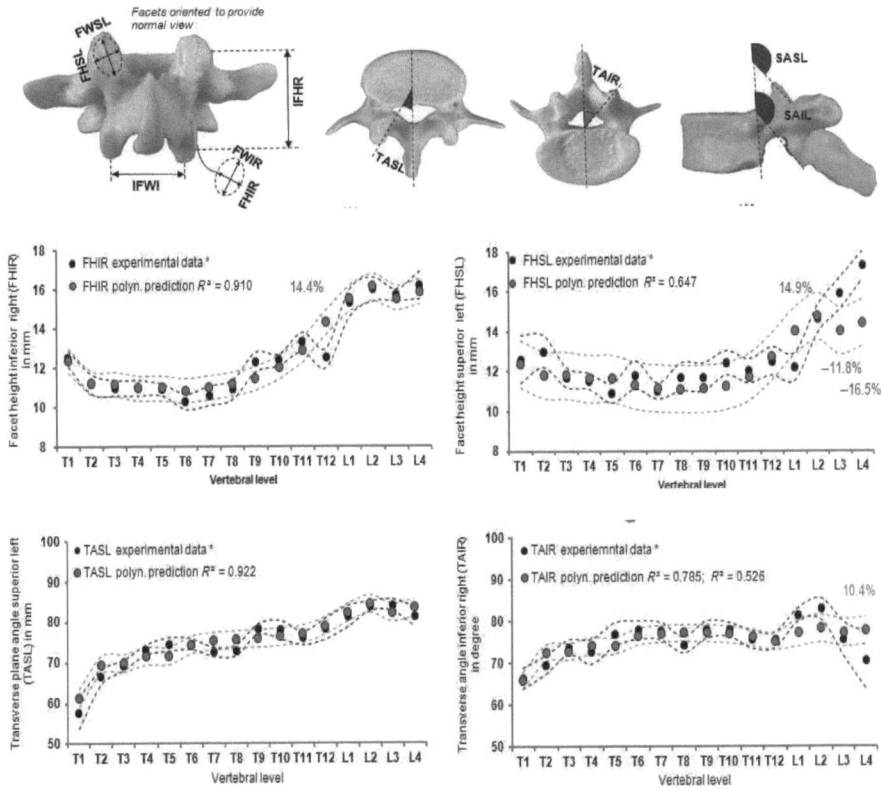

Figure 7: Polynomial predictions of some linear (FHIR and FHSL) and angular (TASL and TAIR) parameters of the thoracic and lumbar articular facet joints from VBHP, superimposed on experimental data of Panjabi et al. (1993)*. Dotted curves indicate standard deviation of both the predictions and experimental data. R^2 is provided for thoracic and for lumbar levels. Only mean percent errors > 5% for all vertebral levels are shown.

Due to the superposition of several anatomical structures, specifically in the sagittal thoracic region of the spine, the direct measurement of the main AFJ parameters considering each vertebral level cannot be performed using lateral radiographs. Moreover, the lumbar AFJ are difficult to image with radiographs because they are both curved and oblique to the sagittal plane. The advantage of using the generated set of prediction equations is the capability to obtain size and orientations of the AFJ considering individual variability from only a single parameter per vertebra (VBHP) measurable on a lateral X-ray. It could be used to provide data for parameterized finite element modelling considering patient-specific AFJ morphometry (Figure 8).

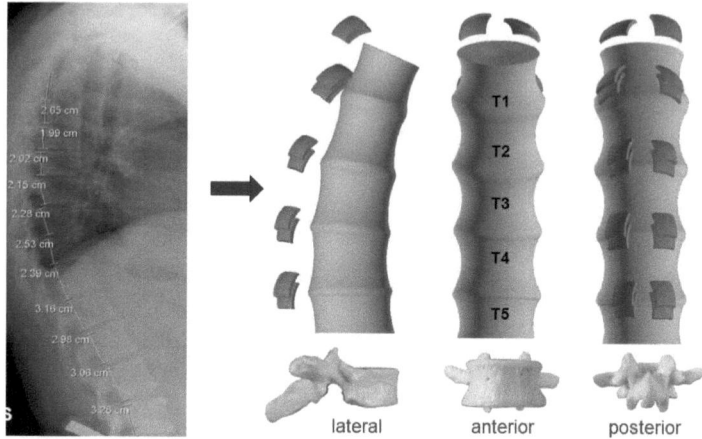

Figure 8: Geometrical model of the thoracic spine including the articular facet joints. The VBHP obtained from lateral radiographs allowed generation of AFJ data using the set of prediction equations.

This study showed that it was possible to establish useful predictors for human thoracic and lumbar AFJ parameters based on the size of the vertebral body. The generated set of prediction equations enables fast acquisition of 20 geometrical parameters of the AFJ as a function of a single parameter (VBHP) which is measurable in lateral radiographs. Since the vertebral body height is unique for each person and vertebral level, the predicted AFJ parameters are also specific to an individual. This approach could be used for parameterized patient-specific modelling of the spine to explore the clinically important mechanical roles of the articular facets in pathological conditions, such as scoliosis.

4. Morphometric analysis of the relationships between intervertebral disc and vertebral body heights: An anatomical and radiographic study of the human thoracic spine

Published in: Kunkel ME, Herkommer A, Reinehr M, Böckers TM, Wilke H-J (2011) Morphometric analysis of the relationships between intervertebral disc and vertebral body heights: An anatomical and radiographic study of the human thoracic spine. *J Anat* **219**, *375-387.*

Morphometric studies on the human intervertebral discs have focused on the cervical and lumbar regions, resulting in limited data on the thoracic region. Intervertebral disc height is an important dimension often used as a diagnostic tool in orthopaedics as well as in mathematical modeling of the human spine. The main aim of this study was to measure the human thoracic intervertebral discs heights by direct and radiographic measurements. Additionally, the heights of the vertebral bodies were measured, and the prediction of the disc heights based on the vertebral bodies was investigated.

Five different heights of the intervertebral discs and vertebral bodies were measured directly and on radiographs of 72 cadaveric spine segments of 15 females (mean age of 58.67 ± 10.74 years, range: 43-80 years) and 15 males (mean age of 56.20 ± 11.65 years, range: 37-79 years) (Figure 9). Six segments were available for each spinal level from C7-T1 to T11-12. A grading system indicated that only mild to moderate degenerative changes were found in the discs and end-plates. Three adimensional morphometric indices were calculated based on studies for lumbar intervertebral discs (Twomey & Taylor 1987; Amonoo-Kuofi 1991).

Lateral radiographs of each spine segment were made using a Faxitron automatic X-ray (Hewlett Packard, Mc Minnville, USA). For the radiographic measurements, individual radiographs were placed on a viewing table and eight anatomical landmarks were identified using Farfan's method (1973) (Figure 9). The disc and vertebral heights were measured using an electronic digital caliper with an accuracy of ± 0.05 mm. Frozen spinal segments were sectioned in the horizontal plane through each of the upper and lower vertebral bodies. A saw microtome was used to produce sagittal sections of the discs (Leica SP4000, Leica Microsystems, Wetzlar, Germany). A sliding vernier caliper was used for the measurement of the disc heights (Mitutoyo, Absolute Digimatic, Tokyo, Japan).

Each set of radiographic and anatomical measurements was carried out by two observers. Inter- and intra-observer errors were examined and expressed as a coefficient of variation (CV). Linear regression was used to examine the correlation between the radiographic and anatomical measurements, and to calculate the accuracy of the radiographic one. The heights of the discs and the vertebral bodies were individually regressed by a least-square estimation process based on Kunkel et al. (2010). Linear and nonlinear regression analyses were employed to find the best functions to fit each of these parameters in a prediction equation.

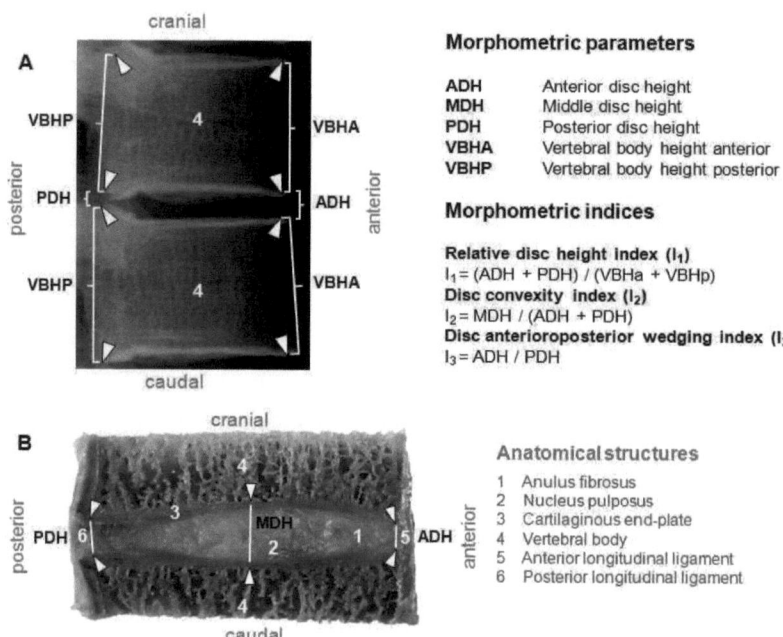

Figure 9: Morphometric parameters and indices of the thoracic intervertebral disc. The anatomical landmarks used are indicated by white arrowheads. In the lateral conventional radiographs of the spinal segments (a) and in the sagittal sections of the specimens (b). The images are from a thoracic segment (T9-10) of a 57 year old female donor.

The parameters ADH, VBHA and VBHP were chosen as predictor variables because they could be measured on the radiographs with an acceptable accuracy. An ANOVA was performed to define the significance of the prediction equations (P-value < 0.05) that were evaluated using experimental data of Todd & Pyle (1928). Radiographic measurements of the disc heights displayed lower repeatability and were shorter than the anatomical ones (approximately 9% for anterior and 37% for posterior heights). The disc height measurements were better repeated when obtained directly from the discs (CV = 0.79 - 0.93) than from the radiographs (CV = 0.49 - 0.82). A lower repeatability was found for radiographic measurement of the PDH (CV= 0.49). Anatomical measurements were reproduced with errors ranging from 1.7% to 6.1% for ADH, 17% to 26.1% for PDH, and 1.7% to 5.1% for VBHA and VBHP. The thickness of the discs varied from 4.5 to 7.2 mm, with the middle heights approximately 22.7% greater. There was a constant relationship between the disc thickness and the vertebral bodies' heights at all levels (ratio disc : body of approximately 1:4.1).

In general, the disc heights showed good correlations with the vertebral body heights (R^2, 0.659-0.835, P-values <0.005) (ANOVA). An exception was the MDH, for which no significant correlations were found (R^2 < 0.6, P-value > 0.05). A set of 10 polynomial equations was generated for the prediction of thoracic disc heights from parameters that could be accurately measured on the radiographs (ADH, VBHA and VBHP). The polynomial predictions were generally within or close to the region of the 95% confidence intervals of the experimental data measured in the current study (Figure 10).

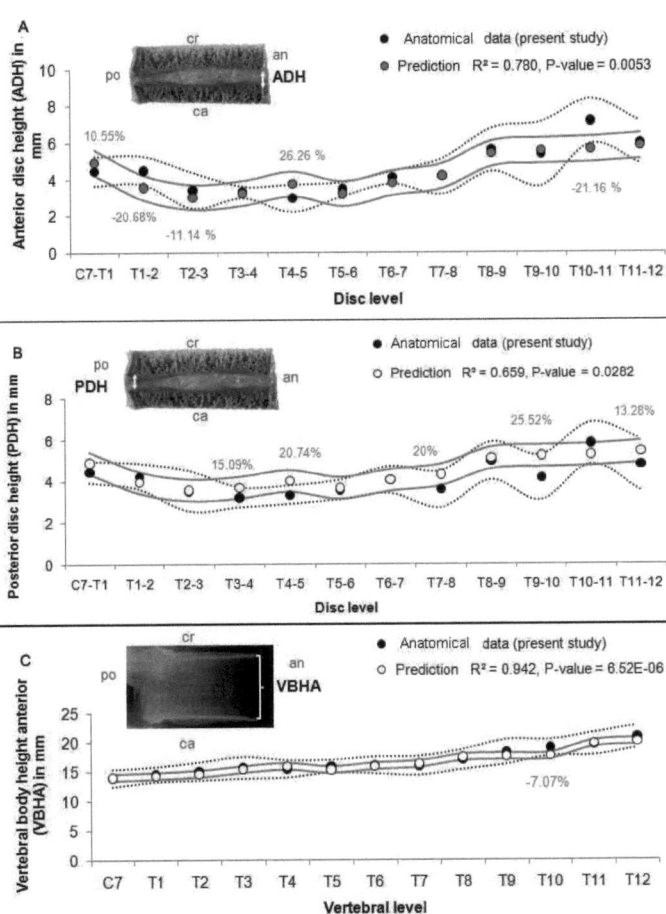

Figure 10: Radiographic values of the VBHP were used for predictions of the parameters ADH, PDH and VBHA at all levels of the thoracic spine. The predicted values were superimposed on experimentally measured anatomical data. Dotted and continuous curves indicate SD. Mean percent errors of the predictions larger than 10% are indicated. Po, posterior; an, anterior; cr, cranial; ca, caudal.

The evaluation of the predictability of the regression equations using VBHA and VBHP of the dataset of the radiographic measurements showed that good results could be obtained. Using the dataset of Todd & Pyle (1928), a comparison of predicted PDH from radiographic ADH showed a greatest error of approximately -13% in the upper and -17% in the lower regions of the thoracic spine.

In the current study, the two main sources of ambiguity found in radiographic measurements of disc heights (the disc orientation with respect to the central X-ray beam, and the estimation of differences among different observers) were minimized using the recommendations of Pope et al. (1977) and Andersson et al. (1981). The difficulty in identifying the bony landmarks was overcome by strictly controlling the vertebral position, and preserving the relationships between the intervertebral discs and the vertebral bodies. The radiographic ADH and PDH measurements were shorter than the anatomical ones probably because the anatomical measurements included the cartilaginous end plates that cannot be readily identified on radiographs. Comparison of these direct and radiographic measurements with other studies was difficult due to the fact that there are few comparative data in the literature (e.g. no published data related to MDH was found). For ADH, a good agreement was found with anatomical values of Todd & Pyle (1928), and radiographic values of Manns et al. (1986) and Giles & Singer (2000), although the same was not found for the radiographic PDH values compared with Giles & Singer (2000).

As expected, the radiographic measurements of the thoracic VBHA and VBHP showed very good agreement with the literature (Todd & Pyle, 1928; Cotterill et al. 1986; Berry et al 1987; Scoles et al. 1988; Panjabi et al. 1991; 1992). The generated set of regression equations allowed prediction of the thoracic disc heights from radiographic measurement of the VBHP. ADH could be predicted, with a largest error of approximately 26% and MDH could only be predicted from ADH (largest error of approximately 15%). For estimation of PDH, both ADH and vertebral heights provided good predictions. From the measurement of the vertebral height were predicted values of PDH with approximately 26% error; which was less than the radiographic measurement. For the creation of parameterized models of the human thoracic discs, the use of these prediction equations could eliminate the need for direct measurement on intervertebral discs reducing the error produced by radiographic measurements for the posterior disc heights (Figure 11).

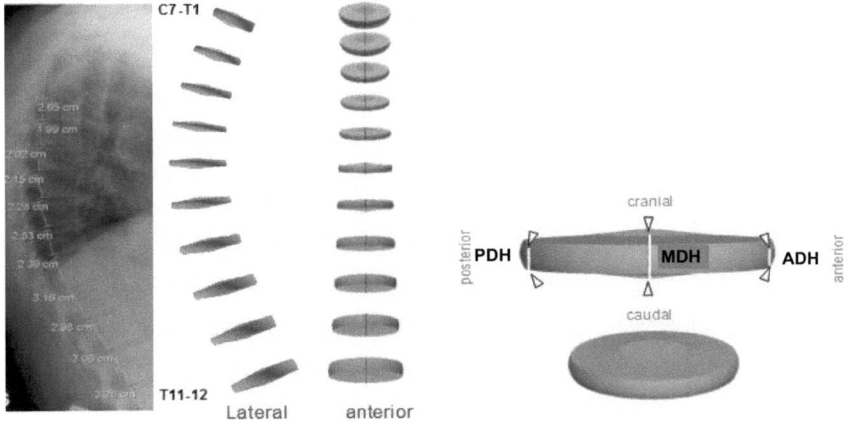

Figure 11: Example of a geometrical model of the thoracic intervertebral discs. The VBHP obtained from lateral radiographs allowed generation of the disc data using the prediction equations.

This study provided an accurate database for the thoracic intervertebral disc heights besides of a statistical approach to predict these parameters from radiographic measurements of the VBHP. This may serve as an anthropometric reference for mathematical modelling as well as for anatomical and biomechanical studies of the human spine.

5. Conclusion

In this work, the morphometric relationships between anatomical data of the human thoracic and lumbar vertebrae and thoracic intervertebral discs were described by linear and nonlinear regression analyses. It allowed the generation and validation of a set of regression equations for the prediction of 40 vertebral and 3 intervertebral disc dimensions per spinal level from the unique measurement of the vertebral body height posterior (VBHP) on lateral radiographs. As the VBHP is unique for each person and vertebral level, the predicted parameters are also specific to an individual. This statistical approach for the acquisition of subject-specific morphometry of the main thoracic and lumbar spinal dimensions from radiographic images may be applied to the construction of parameterized subject-specific models of the spine based on X-ray images alone. Such models allow the performance of studies based on variation of geometry without the need for expensive, invasive and time-consuming data collection, such as direct measurement or reconstruction of medical images.

The hypotheses that were formulated at the beginning of this manuscript were tested and the results showed that: (1) There are only some vertebral dimensions that are measurable on lateral radiographs with ease and accuracy (EPDS, EPDI, VBHA, VBHP and SPL) (Figure 2). No AFJ dimension can be measured on lateral radiographs with ease and accuracy (Figure 6). The intervertebral disc heights cannot be accurately measured on lateral radiographs, the lower repeatability was found for the posterior heights (Figure 9). (2) It was possible to find moderate (e.g. pedicles inclination and transverse process) to high (e.g. the dimensions related to end-plates, spinal canal, heights of pedicles and spinous process) correlation between the investigated dimensions and only one dimension measurable on radiographs, the VBHP. In relation to AFJ parameters, a good correlation was found for the linear dimensions (heights, widths and interfacet distances) and orientations (transverse and sagittal angle) with vertebral body heights. With the exception of the middle disc height, it was possible to find moderate to good correlations for the thoracic intervertebral disc heights and the VBHP. (3) The exclusion of the vertebral dimensions related to the level L5 improved all the correlations. With the exception of the AFJ dimensions, there was not found better correlations between the vertebral dimensions considering only the thoracic or only the lumbar spine. It was found better correlations between the linear and angular AFJ dimensions considering only the thoracic levels. Correlations considering only lumbar data on AFJ did not satisfy the criteria required and for this reason correlations using the combined thoracic and lumbar data together were used. (4) The analysis of covariance indicated that nonlinear regressions (third order polynomial) were able to describe these correlations better than linear regressions. (5) Using the measurement of only the VBHP and the generated set of prediction equations, it was possible to predict vertebral morphometry of the thoracic and lumbar spine with the exception of the level L5. For AFJ dimensions, the best predictions were achieved for the thoracic levels that showed a mean error of approximately 10%, while the lumbar level displayed a mean error of approximately 16% (ANOVA). Using the measurement of only the VBHP and the generated set of prediction equations, it was possible to predict thoracic intervertebral disc heights with an error less than the error of direct measurement. (6) Many vertebral dimensions that are not visible and that cannot be measured on lateral radiographs could be predicted with this statistical approach (e.g. EPWS, EPWI, PWL, PWRPSI, PSIL, SCW, SCD and TPW). All the selected AFJ dimensions that are not visible and that cannot be measured on lateral radiographs could also be predicted. The middle intervertebral disc height of the thoracic spine that does not show a good visibility on lateral radiographs could also be predicted. (7) The radiographic measurement of the VBHP in the thoracic and lumbar levels of the spine of a subject allowed specific predictions for this subject. (8) The regression equations could be validated for some vertebral dimensions by comparisons of predictions with experimental data. For the thoracic intervertebral discs, the regression equations could only be validated for the posterior disc heights by comparisons of predictions with experimental data.

6. Summary

Severe cases of scoliosis are treated using implants. The effects of different types of implants have been investigated with mathematical models with modifiable geometry. Our aim was perform a nonlinear regression analyses with anatomical data to generate prediction equations for vertebral and intervertebral disc dimensions as a function of only one given dimension measurable by X-ray, the vertebral body height. Third-order polynomial regressions provided moderate to high correlation between the vertebral body heights and the endplates and spinal canal; pedicle heights and the spinous process, in addition to a reasonable correlation of the posterior vertebral structures (pedicle and facet). A set of 50 equations was generated for the prediction of the spine dimensions based on the radiographic measurement of the vertebral body height. It was possible to establish useful predictions for all investigated dimensions. This is an efficient approach for obtaining anatomical data for modeling of the human thoracic and lumbar from measurement of only one parameter per vertebra without the need for direct measurement or 3D reconstructions from medical images.

7. References

Abumi K, Panjabi MM, Kramer KM, Duranceau J, Oxland T, Crisco JJ (1990) Biomechanical evaluation of lumbar spinal stability after graded facetectomies. *Spine* **15**, 1142-1147.

Abuzayed B, Tutunculer B, Kucukyuruk B, Tuzgen S (2010) Anatomic basis of anterior and posterior instrumentation of the spine: morphometric study. *Surg Radiol Anat* **32**, 75-85.

Adams MA, Hutton WC (1983) The mechanical function of the lumbar apophyseal joints. *Spine* **8**, 327-330.

Aebi M (2005) The adult scoliosis. *Eur Spine J* **14**, 925-948.

Aharinejad S, Bertagnoli R, Wicke K, Firbas W, Schneider B (1990) Morphometric analysis of vertebrae and intervertebral discs as a basis of disc replacement. *Am J Amat* **189**, 69-76.

Ahmed AM, Duncan NA, Burke DL (1990) The effect of facet geometry on the axial torque-rotation response of lumbar motion segments. *Spine* **15**, 391-401.

Amonoo-Kuofi HS (1991) Morphometric changes in the heights and anteroposterior diameters of the lumbar intervertebral disc with age. *J Anat* **175**, 159-168.

Andersson GBJ, Schultz A, Nathan A, Irstam L (1981) Roentgenographic measurement of lumbar intervertebral disc height. *Spine* **6**, 154-157.

Aubin C-É, Dansereau J, Parent F, Habelle H, De Guise JA (1997) Morphometric evaluation of personalised 3D reconstructions and geometrical models of the human spine. *Med Biol Eng Comput* **35**, 611-618.

Benameur S, Mignotte M, Labelle H, de Guise (2005) A hierarchical statistical modeling approach for the unsupervised 3-D biplanar reconstruction of the scoliotic spine. *IEEE T Bio-Med Eng* **52**, 2041-2057.

Berry JL, Moran JM, Berg WS, Steffee AD (1987) A morphometric study of human lumbar and selected thoracic vertebrae. *Spine* **12**, 362-367.

Breglia DP (2006) *Generation of a 3-D parametric solid model of the human spine using anthropomorphic parameters.* Master Dissertation, Ohio University, Ohio.

Boszczyk BM (1997) *Wirbel und Bewegung – Vergleichende Anatomie der Lendenwirbel – speziell der Wirbelgelenke und Versuch einer Kausalen Analyse.* PhD Dissertation, University of Munchen, Munchen [in German].

Carman DL, Browne RH, Birch JG (1990) Measurement of scoliosis and kyphosis radiographs. Intraobserver and interobserver variation. *J Bone Joint Surg Am* **72**, 328-333.

Cheriet F, Dansereau J, Petit Y, Aubin C-É, Labelle H, De Guise JA (1999a) Towards the self-calibration of a multiview radiographic imaging system for the 3d reconstruction of the human spine and rib cage. *Int J of Pattern Recogn* **13**, 761-779.

Cheriet F, Delorme S, Dansereau J (1999b) Intraoperative 3D reconstruction of the scoliotic spine from radiographs. *Ann Chir* **53**, 808-815.

Cotterill PC, Kostuik JP, D'Angelo GD, Fernie GR, Maki BE (1986) An anatomical comparison of the human and bovine thoracolumbar spine. *J Orthop Res* **4**, 298-303.

Dai LY (2001) Orientation and tropism of lumbar facet joints in degenerative spondylolisthesis. *Int Orthop* **25**, 40-42.

Dansereau J, Stokes I (1988) Measurement of the three-dimensional shape of the rib cage. *J Biomech* **21**, 893-901.

Duke K, Aubin C-É, Dansereau J, Labelle H (2004) Biomechanical simulations of scoliotic spine correction due to prone position and anaesthesia prior to surgical instrumentation. *Clin Biomech* **20**, 923-931.

Dumas R, Mitton D, Laporte S, Dubousset J, Steib JP, Lavaste F, Skalli W (2003) Explicit calibration method and specific device designed for stereoradiography. *J Biomech* **36**, 827-834.

Dumas R, Le Bras A, Champain N, Savidan M, Mitton D, Kalifa G, Steib J-P, De Guise JA, Skalli W (2004) Validation of the relative 3D orientation of vertebrae reconstructed by bi-planar radiography. *Med Eng Phys* **26**, 415-422.

Dumas R, Lafage V, Lafon Y, Steip J-P, Mitton D, Skalli W (2005) Finite element simulation of spinal deformities correction by in situ contouring technique. *Comput Method Biomec* **8**, 331-337.

Dumas R, Blanchard B, Cartier R, de Loubresse CG, Le Huec J-C, Marty C, Moinard M, Vital J-M (2008) A semi-automated method using interpolation and optimisation for the 3D reconstruction of the spine from bi-planar radiography: A precision and accuracy study. *Med Biol Eg Comp* **46**, 85-92.

Duncan NA, Ahmed AM (1989) The influence of the apophyseal joints on the 3-D kinematics of lumbar motion segments. *ASME Adv Bioeng*, 81-82.

Dupuis PR, Yong-Hing K, Cassidy JD, Kirkaldy-Willis WH (1985) Radiologic diagnosis of degenerative lumbar spinal instability. *Spine* **10**, 262-276.

Ebraheim NA, Xu R, Ahmad M, Yeasting RA (1997) The quantitative anatomy of the thoracic facet and the posterior projection of its inferior facet. *Spine* **22**, 1811-1817.

Eijkelkamp MF (2002) *On the development of an artificial intervertebral disc*. PhD-Dissertation, University of Groningen.

Eisenstein S (1976) Measurements of the lumbar spinal canal in 2 racial groups. *Clin Orthop* **115**, 42-45.

Eisenstein S (1977) The morphometry and pathological anatomy of the lumbar spine in south african negroes and caucasoids with specific reference to spinal stenosis. *J Bone and Joint Surg* **59**, 173-180.

Eisenstein S (1983) Lumbar vertebral canal morphometry for computerised tomography in spinal stenosis. *Spine* **8**, 187-191.

Farfan HF, Sullivan JD (1967) The relation of facet orientation to intervertebral disc failure. *Canadian J Surgery* **10**, 179-185.

Farfan HF, Huberdeau RM, Dubow HI (1972) The influence of geometrical features on the pattern of disc degeneration – A post mortem study. *J Bone Joint Surg* **54**, 492-510.

Frederick Harrington J Jr, Sungarian A, Rogg J, James Makker V, Epstein MH (2001) The relation between vertebral endplate shape and lumbar disc herniations. *Spine* **26**, 2133-2138.

Gilad I, Nissan M (1986) A study of vertebra and disc geometric relations of the human cervical and lumbar spine. *Spine* **11**, 154-157.

Gilbertson LG, Goel VK, Kong WZ (1995) Finite element methods in spine biomechanics research. *Crit Rev Biomed Eng* **23**, 411-73.

Giles LGF, Singer KP (2000) *Clinical anatomy and management of thoracic spine pain*, pp.25-27. Oxford: Butterworth-Heinemann.

Goel VK, Gilbertson LG (1995) Spine Update – Applications of the finite element method to thoracolumbar spinal research – Past, present, and future. *Spine* **20**, 1719-1727.

Goel VK, Weinstein JN (1990) *Biomechanics of the Spine: Clinical and Surgical Perspective*, pp.14-15. Florida: CRC Press.

Goh S, Price RI, Leedman PJ, Singer KP (1999) The relative influence of vertebral body and intervertebral disc shape on thoracic kyphosis. *Clin Biomech* **14**, 439-448.

Hall LT, Esses SI, Noble PC (1998) Morphology of the lumbar vertebral endplates. *Spine* **23**, 1517-1522.

Hansson TH, Hansson EK (2000) The effects of common medical interventions on pain, back function, and work resumption in patients with chronic low back pain: A prospective 2-year cohort study in six countries. *Spine* **25**, 3055-3064.

van der Houwen EB, Baron P, Veldhuizen AG, Burgerhof JGM, van Ooijen PMA, Verkerke GJ (2010) Geometry of the intervertebral volume and vertebral endplates of the human spine. *Ann Biomed Eng* **38**, 33-40.

Huizinga J, van der Heiden jA, Finken PJJG (1952) The human lumbar vertebral canal: A biometric study. *Proc Roy Netherlands Acad Sci* **C55**, 22-23.

Hurxthal LM (1968) Measurement of anterior vertebral compressions and biconcave vertebrae. *Am J Roentgenol Rad Ther Nucl Med* **103**, 635-644.

Jason DR, Tyalor K (1995) Estimation of stature from the length of the cervical, thoracic, and lumbar segments of the spine in American whites and blacks. *J Forensic Sci* **40**, 59-62.

Jones RAC, Thomson JLG (1968) The narrow lumbar canal. A clinical and radiological review. *J Bone Joint Surg Br* **50**, 595-609.

Kadoury S, Cheriet F, Dansereau J (2007a) Three-dimensional reconstruction of the scoliotic spine and pelvis from uncalibrated biplanar X-ray images. *J Spinal disord Tech* **20**, 160-167.

Kadoury S, Cheriet F, Laporte C (2007b) A versatile 3d reconstruction system of the spine and pelvis for clinical assessment of spinal deformities. *Med Biol Eng Comput* **45**, 591-602.

Kósa F, Castellana C (2005) New forensic anthropological approachment for the age determination of human fetal skeletons on the base of morphometry of vertebral column. *Forensic Sci Int* **147**, S69-S74.

Krag MH, Beynnon BD, Pope MH, Frymoyer JW, Haugh LD, Weaver DL, et al. (1986) An internal fixator for posterior application to short segments of the thoracic, lumbar, or lumbosacral spine: Design and testing. *Clin Orthop* **203**, 75-98.

Krag MH, Weaver DL, Beynnon BD, Haugh LD (1988) Morphometry of the thoracic and lumbar spine related to transpedicular screw placement for surgical spinal fixation. *Spine* **13**, 27-32.

Kunkel ME, Schmidt H, Wilke H-J (2010) Prediction equations for human thoracic and lumbar vertebral morphometry. *J Anat* **216**, 320-328.

Kunkel ME, Schmidt H, Wilke H-J (2011a) Prediction of the human thoracic and lumbar articular facet joint morphometry from radiographic images. *J Anat* **218**, 191-201.

Kunkel ME, Herkommer A, Reinehr M, Böckers TM, Wilke H-J (2011b) Morphometric analysis of the relationships between intervertebral discs and vertebral body heights: An anatomical and radiographic study of the human thoracic spine. *J Anat* **219**, 375-387.

Lafage V, Dubousset J, Lavaste F, Skalli W (2002) Finite element simulation of various strategies for CD correction. *Stud Health Technol Inform* **91**, 428-432.

Lafage V, Dubousset J, Lavaste F, Skalli W (2004) 3D finite element simulation of Cotrel-Dubousset correction. *Comput Aided Surg* **9**, 17-25.

Laporte S, Mitton D, Ismael B, de Fouchecour M, Lasseau JP, Lavaste F, Skalli W (2000) Quantitative morphometric study of thoracic spine. A preliminary parameters statistical analysis. *Eur J Orthop Surg Traumatol* **10**, 85-91.

Larsen JL (1985a) The posterior surface of the lumbar vertebral bodies. Part I. *Spine* **10**, 50-58.

Larsen JL (1985b) The posterior surface of the lumbar vertebral bodies. Part II: An anatomic investigation concerning the curvatures in the horizontal plane. *Spine* **10**, 901-906.

Lavaste F, Skalli W, Robin S, Roy-Camille R, Mazel C (1992) Three-dimensional geometrical and mechanical modelling of the lumbar spine. *J Biomech* **25**, 1153-1164.

Lee HM, Kim NH, Kim HJ, Chung I-H (1995) Morphometric study of the lumbar spinal canal in the Korean population. *Spine* **20**, 1679-1684.

Lilijenqvist UR, Link TM, Halm HFH (2000) Morphometric analysis of thoracic and lumbar in idiopathic scoliosis. *Spine* **25**, 1247-1253.

Lorenz M, Patwardhan A, Vanderby R (1983) Load-bering characteristics of lumbar facets in normal and surgically altered spinal segments. *Spine* **8**, 122-130.

Manns RA, Haddaway MJ, McCall IW, Pullicino VC, Davie MWJ (1996) The relative contribution of disc and vertebral morphometry to the angle of kyphosis in asymptomatic subjects. *Clin Radiol* **51** 258-262.

Marchesi D, Schneider E, Glauser P, Aebi M (1988) Morphometric analysis of the thoracolumbar and lumbar pedicles, anatomo-radiologic study. *Surg Radiol Anat* **10**, 317-322.

Masharawi Y, Rothschild B, Dar G, Peleg S, Robinson D, Been E, Hershkovitz I (2004) Facet orientation in the thoracolumbar spine: Three-dimensional anatomic and biomechanical analysis. *Spine* **29**, 1755-1763.

Masharawi Y, Rothschild B, Salame K, Dar G, Peleg S, Hershkovitz I (2005) Facet tropism and interfacet shape in the thoracolumbar vertebrae. *Spine* **30**, E281-E292.

Masharawi Y, Dar G, Peleg S, Steinberg N, Alperovitch-Najenson D, Salame K, Hershkovitz I (2007a) Lumbar facet anatomy changes in spondylolysis: A comparative skeletal study. *Eur Spine J* **16**, 993-999.

Masharawi YM, Alperovitch-Najenson D, Steinberg N, Dar G, Peleg S, Rothschild B, Salame K, Hershkovitz I (2007b) Lumbar facet orientation in spondylolysis: A skeletal study. *Spine* **32**, E176-E180.

Maurel N, Lavaste F, Skalli W (1997) A three-dimensional parameterized finite element model of the lower cervical spine. Study of the influence of the posterior articular facets. *J Biomech* **30**, 921-931.

McLain RF, Yerby SA (2002) Comparative morphometry of L4 vertebrae: Comparison of large animal models for the human lumbar spine. *Spine* **27**, E200-E2006.

Millner PA, Dickson RA (1996) Idiopathic scoliosis: biomechanics and biology. *Eur Spine J* **5**, 362-373.

Moran JM, Berg WS, Berry JL, Geiger JM, Steffee AD (1989) Transpedicular screw fixation. *J Orthop Res* **7**, 107-114.

Nissan M, Gilad I (1984) The cervical and lumbar vertebrae - An anthropometric model. *Engin Med* **13**, 111-114.

Nissan M, Gilad I (1986) Dimensions of the human lumbar vertebrae in the sagittal plane. *J Biomech* **19**, 743–758.

Nojiri K, Matsumoto M, Chiba K, Toyama Y (1990) Morphometric analysis of the thoracic and lumbar spine in Japanese on the use of pedicle screws. *Surg Radiol Anat* **27**, 123-128.

O'Connell GD, Vresilovic EJ, Elliott DM (2007) Comparison of animals used in disc research to human lumbar disc geometry. *Spine* **32**, 328-333.

Olsewski JM, Simmons EH, Kallen FC, Mendel FC, Severin CM, Berens DL (1990) Morphometry of the lumbar spine: Anatomical perspectives related to transpedicular fixation. *J Bone and Joint Surg* **72**, 541-548.

Onan OA, Hipp JA, Heggeness MH (1998) Use of computed tomography image processing for mapping of human cervical facet surface geometry. *Med Eng Phys* **20**, 77-81.

Panjabi MM, Yamamoto I, Oxland T, Crisco J (1989) How does posture affect coupling in the lumbar spine? *Spine* **14**, 1002-1011.

Panjabi MM, Takata K, Goel V, Federico D, Oxland T, Duranceau J, Krag M (1991) Thoracic human vertebrae - Quantitative three-dimensional anatomy. *Spine* **16**, 888-901.

Panjabi MM, Goel V, Oxland T, Takata K, Duranceau J, Krag M, Price M (1992) Human lumbar vertebrae - Quantitative three-dimensional anatomy. *Spine* **17**, 299-306.

Panjabi MM, Oxland T, Takata K, Goel V, Duranceau J, Krag M (1993) Articular facets of the human Spine - quantitative three-dimensional anatomy. *Spine* **18**, 1298-1310.

Parent S, Labelle H, Skalli W, Latimer B, de Guise J (2002) Morphometric analysis of anatomic scoliotic specimens. *Spine* **27**, 2305-2311.

Parent S, Labelle H, Skalli W, de Guise J (2004) Thoracic pedicle morphometriy in vertebrae from scoliotic spines. *Spine* **29**, 239-248.

Petiti Y, Dansereau J, Labelle H, de Guise JA (1998) Estimation of 3D location and orientation of human vertebral facet joints from standing digital radiographs. *Med Biol Eng Comput* **36**, 389-394.

Pomero V, Mitton D, Laporte S, de Guise JA (2004) Fast accurate stereographic 3D-reconstruction of the spine using a combined geometric and statistic model. *Clin Biomech* **19**, 240-247.

Pooni JS, Hukins DWL, Harris PF, Hilton RC, Davies KE (1986) Comparison of the structure of human intervertebral discs in the cervical, thoracic and lumbar regions of the spine. *Surg Radiol Anat* **8**, 175-182.

Pope MH, Hanley EN, Matteri RE, Wilder D, Frymoyer JW (1977) Measurement of intervertebral disc space height. *Spine* **2**, 282-286.

Porter RW, Hibbert C, Wellman P (1980) Backache and the lumbar spinal canal. *Spine* **5**, 99-105.

Postacchini F, Ripani M, Carpano S (1983) Morphometry of the lumbar vertebrae. An anatomic study in two caucasoid ethnic groups. *Clin Orthop* **172**, 296-303.

Robin, S, Skalli W, Lavaste, FE (1994) Influence of geometrical factors on the behavior of lumbar spine segments: a finite element analysis. *Eur Spine J* **3**, 84-90.

van Schaik JPJ, Verbiest H, Frans DJ (1985a) The orientation of laminae and facet joints in the lower lumbar spine. *Spine* **10**, 59-63.

van Schaik JJ, Verbiest H, van Schaik FDJ (1985b) Morphometry of lower lumbar vertebrae as seen on CT scans: Newly recognized characteristics. *Am J Roentgenol* **145**, 327-335.

Schmidt H, Heuer F, Wilke H-J (2008a) Interaction between finite helical axes and facet joint forces under combined loading. *Spine* **33**, 2741-2748.

Schmidt H, Heuer F, Claes L, Wilke H-J. (2008b) The relation between the instantaneous center of rotation and facet joint forces - A finite element analysis. *Clin Biomech* **23**, 270-278.

Schmidt H, Midderhoff S, Adkins K, Wilke H-J (2009) The effect of different design concepts in lumbar total disc arthroplasty on the range of motion, facet joint forces and instantaneous center of rotation of a L4-5 segment. *Eur Spine J* **18**, 1695-1705.

Scoles PV, Linton AE, Latimer B, Levy ME, Digiovanni BF (1988) Vertebral body and posterior element morphology: the normal spine in middle life. *Spine* **13**, 1082-1086.

Semaan I, Skalli W, Veron S, Templier A, Lassau JP, Lavaste F (2001) Quantitative 3D anatomy of the lumbar spine. *Rev Chir Orthop Reparatrice Appar Mot* **87(4)**, 340-53. [in French]

Shao Z, Rompe G, Schiltenwolf M (2002) Radiographic changes in the lumbar intervertebral disc and lumbar vertebrae with age. *Spine* **27**, 263-268.

Shirazi-Adl A (1991) Finite-element evaluation of contact loads on facets of an L2-L3 lumbar segment in complex loads. *Spine* **16**, 533-541.

Shirazi-Adl A (1994) Nonlinear stress analysis of the whole lumbar spine in torsion – mechanics of facet articulation. *J Biomech* **27**, 289-299.

Singh R, Srivastva SK, Prasath CSV, Rohilla RK, Siwach R, Magu NK (2011) Morphometric measurements of cadaveric thoracic spine in indian population and its clinical applications. *Asian Spine J* **5**, 20-34.

Skalli W, Robin SI, Lavaste F, Dubousset J (1993) A biomechanical analysis of short segment spinal fixation using a three-dimensional geometric and mechanical model. *Spine* **18**, 536-545.

Stokes IAF (**1994**) Three-dimensional terminology of spinal deformity: A report presented to the scoliosis research society by the scoliosis research society working group on 3-D terminology of spinal deformity. *Spine* **19**, 123-256.

Taylor JR, Twomey LT (1986) Age changes in lumbar zygapophyseal joints: Observations on structure and function. *Spine* **11**, 739-745.

Tibrewal SB, Pearcy MJ (1985) Lumbar intervertebral disc heights in normal subjects and patient with disc herniation. *Spine* **10**, 452-454.

Todd TW, Pyle SI (1928) A quantitative study of the vertebral column by direct and roentgenoscopic methods. *Am J Phys Anthropol* **XII**, 321-338.

Twomey L, Taylor JR (1987) Degenerative age changes in lumbar vertebrae and intervertebral discs. *Clin Orthop* **224**, 97-104.

Wang J, Yang X (2009) Age-related changes in the orientation of lumbar facet joints. *Spine* **34**, E596-E598.

Wenig CM, Schmidt CO, Kohlmann T, Schweikert B (2009) Costs of back pain in Germany. *Eur J Pain* **13**, 280-286.

White AA, Panjabi MM (1990) *Clinical Biomechanics of the Spine,* pp. 39-40. Philadelphia: Lippincott.

Wilke H-J, Krischak S; Claes LE (1996) Biomechanical comparison of calf and human spines. *J Orthopaed Res* **14**, 500-5003.

Wilke H-J, Kettler A, Wenger KH, Claes LE (1997b) Anatomy of the sheep spine and its comparison to the human spine. *Anat Record* **247**, 542-555.

Wilke H-J, Kettler A; Claes LE (1997b) Are sheep spines a valid biomechanical model for human spines? *Spine* **22**, 2365-2374.

Yazici M, Acaroglu E, Alanay A, Deviren V, Cila A, Surat A (2001) Measurement of vertebral rotation in standing versus supine position in adolescent idiopathic scoliosis. *J Pediatr Orthoped* **21**, 252-256.

Yu S-B, Lee U-Y (2008) Determination of sex for the 12th thoracic vertebra by morphometry of three-dimensional reconstructed vertebral models. *J Forensic Sci* **53**, 620-625.

Zander T, Rohlmann, Klöckner C, Bergmann G (2003) Influence of graded facectomy and laminectomy on spinal biomechancis. *Eur Spine J* **12**, 427-434.

Zindrick MR, Wiltse LL, Doornik A, Widell EH, Knight GW, Patwardhan AG, Thomas JC, Rothman SL, Fields BT (1987) Analysis of the morphometric characteristics of the thoracic and lumbar pedicles. *Spine* **12**, 162-166.

8. Papers

Prediction equations for human thoracic and lumbar vertebral morphometry

Maria E. Kunkel, Hendrik Schmidt and Hans-Joachim Wilke

Institute of Orthopaedic Research and Biomechanics, University of Ulm, Ulm, Germany

Abstract

Statistical correlations between anatomical dimensions of human vertebral structures have indicated a potential for the prediction of vertebral morphometry, which could be applied to the creation of simplified geometrical models of the spine excluding the need for preliminary processing of medical images. The aim of this study was to perform linear and nonlinear regressions with published anatomical data to generate prediction equations for 20 vertebral parameters of the human thoracic and lumbar spine as a function of only one given parameter that was measured by X-ray. Each parameter was considered individually as a potential predictor variable in terms of its correlation with all of the other parameters, together with the readiness with which lateral X-rays could be obtained. Based on this, the parameter vertebral body height posterior was chosen and the statistical analyses described here are related to this parameter. Our linear, exponential and logarithmic regressions provided significant predictions of anterior vertebral structures. However, third-order polynomial prediction equations allowed an improvement on these predictions (P-values < 0.001), e.g. endplates and spinal canal (R^2, 0.970–0.995) as well as pedicle heights and the spinous process (R^2, 0.811–0.882), in addition to a reasonable prediction of the posterior vertebral structures, which have shown a low or no correlation in previous studies, e.g. pedicle inclination and transverse process (R^2, 0.514–0.693) (ANOVA). Comparisons of the theoretical predictions with two other sets of experimental data indicated that the predictions generally agree well with the experimental data. A time-efficient approach for obtaining anatomical data for the description of human thoracic and lumbar geometry was provided by this method, which requires the measurement of only one parameter per vertebra (vertebral body height posterior) from a lateral X-ray and the set of developed prediction equations. Vertebral models based on this type of parameterized geometry could be used in biomechanical studies that require geometry variation, such as in spinal deformations, including scoliosis.

Key words polynomial regression; prediction; vertebral morphometry; vertebral parameters.

Introduction

During recent decades, finite element analyses have been performed to provide a better understanding of the biomechanics of the human spine. Several finite element models have been developed and are summarized in Gilbertson et al. (1995) and Fagan et al. (2002). As geometrical factors exert a noticeable influence on the behavior of the spine (Robin et al. 1994), reliable simulations of human spine behavior require complex 3D modeling of the main anatomical structures, e.g. vertebrae, intervertebral discs and ligaments.

Correspondence
Hans-Joachim Wilke, Institute of Orthopaedic Research and Biomechanics, Helmholtzstrasse 14, D-89081 Ulm, Germany. T: 0049 731 500 55320; fax: 0049 731 500 55302; E: hans-joachim.wilke@uni-ulm.de

Accepted for publication *3 November 2009*
Article published online *21 December 2009*

Human vertebral geometry has typically been obtained, *in vivo*, through the 3D reconstruction of medical images, such as computed tomography or magnetic resonance imaging. This technique provides accurate vertebral assessment but requires a long processing time and considerable computational power is required for the manual or semi-automatic segmentation of the images. Moreover, the patient has to be submitted to relatively high doses of ionizing radiation. Alternative procedures have included stereo-radiographic approaches using X-rays (Aubin et al. 1997; Dumas et al. 2004). However, these require a long and tedious process of identification of numerous anatomical landmarks. Some semi-automatic methods have shown fast vertebral reconstruction (Pomero et al. 2004) but they require specific software and hardware.

In-vitro measurements with cadaveric vertebrae have been taken directly from bony specimens or have been obtained from medical images (Krag et al. 1988). These studies have focused on only one specific anatomic structure, such as the dimensions of the vertebral body (Hall et al. 1998), spinal canal, pedicles (Zindrick et al. 1987; Marchesi et al. 1988; Moran et al. 1989) and facet joints (Masharawi et al. 2004); a limited set of structures (Berry et al. 1987); or a limited set of vertebrae such as thoracic (Cotterill et al. 1986; Scoles et al. 1988; Aharinejad et al. 1990) or lumbar vertebrae (Semaan et al. 2001). The most complete collection of quantitative 3D-surface anatomy of the main vertebral parameters for the thoracic and lumbar human spine has been provided in Panjabi et al. (1991, 1992). As this dataset has been used in the current study, a detailed description of the measured parameters is provided in the Materials and methods.

Investigations of correlations between anatomical dimensions of the human vertebral structures have indicated that vertebral relationships could be used to predict vertebral morphometry without the preliminary processing of medical images. Statistical analyses that were performed using simple linear regression analyses between the main vertebral parameters and the vertebral body height (e.g. Scoles et al. 1988; Breglia, 2006) have found low or no correlations for some important parameters, such as pedicle dimensions. Scoles et al. (1988) described the posterior structures as being highly variable and largely unpredictable.

X-rays are frequently used in clinical diagnosis for patients as well as in biomechanical experiments with human vertebral samples. Some studies have used multiple-linear regression analyses (e.g. Lavaste et al. 1992; Laporte et al. 2000) to provide methods for the reconstruction of the human vertebrae from two X-rays (anterior–posterior and lateral). However, to explain 100% of the variability for each parameter, the measurement of six to 15 initial parameters per vertebra on X-rays was needed. Moreover, none of these previous studies have performed an evaluation of the predictability of the generated equations with another set of experimental measurements.

The aim of this study was to perform linear and nonlinear regression analyses with published anatomical data to generate prediction equations for 20 vertebral parameters of the human thoracic and lumbar spine as a function of only one given parameter measured by X-ray.

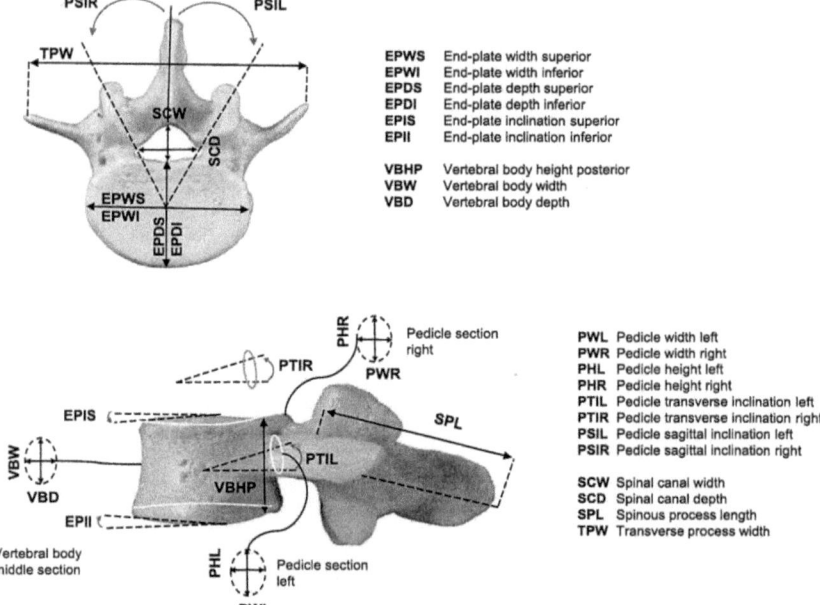

Fig. 1 Schematic representation of the vertebral anatomical parameters that were considered for linear and nonlinear regression analyses.

© 2009 The Authors
Journal compilation © 2009 Anatomical Society of Great Britain and Ireland

Materials and methods

Study population

Vertebral anatomical data were collected from the studies of Panjabi et al. (1991, 1992) and included in this study. This dataset was considered as being an approximate average for non-pathological human spines. It provided linear and angular dimensions of the main parameters from human cadaveric thoracic and lumbar vertebrae. The mean age of the subjects ($n = 12$) was 46.3 years (range: 19–59 years), weight was 67.8 kg (range: 54–85 kg), height was 167.8 cm (range: 157–178 cm) and the male : female ratio was 8 : 4. In order to carry out statistical analyses, 15 linear and six angular vertebral parameters were selected from this dataset to describe the size and shape of the human thoracic (T1–12) and lumbar (L1–4) vertebrae (Fig. 1). The values of the vertebral parameters related to the vertebral level L5 were not included in the analysis.

Statistical analysis

The initial assumption for this study was that 20 vertebral parameters on each level of the thoracic (T1–12) and lumbar (L1–4) spine could be considered as a variable that can be predicted (Fig. 1). All vertebral parameters that were selected for this study were tested as a possible predictor variable. Each vertebral parameter was individually regressed against the possible predictor variable by a least-squares estimation process. Based on the level of correlation with the other parameters and ease of measurement on lateral X-rays, the parameter vertebral body height posterior (VBHP) was chosen and the statistical analyses described in this study are related to this parameter. Linear and nonlinear regression analyses were employed to find the best functions to fit each parameter in a prediction equation.

During the statistical analyses, several hypotheses were tested for each parameter: (i) a function could not fit the data significantly better than a horizontal line (no relationship between the two selected variables); (ii) a second-order equation could not fit the data significantly better than a linear equation; (iii) a third-order equation could not fit the data significantly better than a second-order equation, and so on. The statistical procedure performed on each parameter corresponds to a four-step procedure that is illustrated as an example in Fig. 2.

In the first step, least-squares estimation was used to find equations to describe the relationship between two vertebral parameters, e.g. the variables (EPWS) and VBHP; each data point represents the mean value in different vertebral levels of 12 cadavers. Initially, a linear regression was performed, fitting an equation of the form $y = C_1 + C_2 x$ to the data. The fraction of the overall variance of the EPWS that was reduced by this line was determined by the R^2 value. A logarithmic and an exponential curve with equations of the form $y = C_1 + C_2 \ln(x)$ and

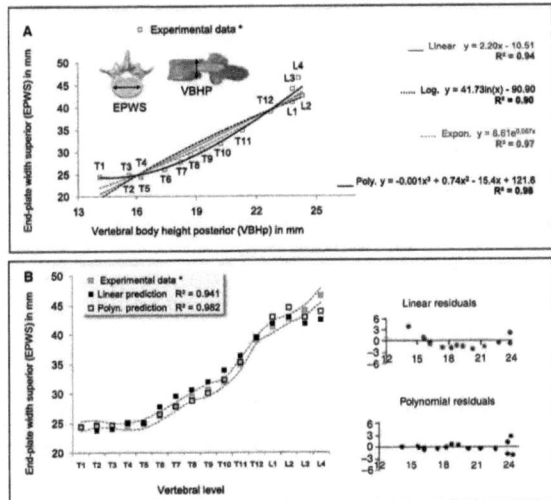

Fig. 2 Description of the statistical procedure performed for two vertebral parameters. (A) Correlation of experimental data of the thoracic and lumbar spine (in this case, VBHP vs. EPWS) and set of prediction equations generated from linear, logarithmic, exponential and polynomial regression analyses (y is the value of EPWS and x is the value of VBHP on each vertebral level). (B) Values of EPWS predicted using linear and polynomial equations are superimposed on experimental data to allow the selection of the best equation. Dotted curve indicates SD of the experimental data. Residual plots evaluation shows that the polynomial equation is significantly better.

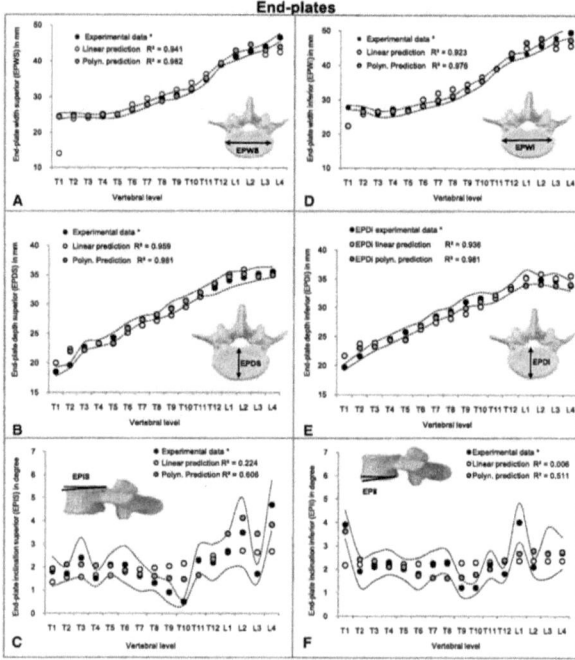

Fig. 3 Linear and polynomial predictions of parameters related to endplates and vertebral body (EPWS, EPWI, EPDS, EPDI, EPIS and EPII) (A-F) superimposed on experimental data of Panjabi et al. (1991, 1992)*. Dotted curve indicates SD of the experimental data.

$y = C_1 e^{C_2 x}$, respectively, were then used to test the increase in R^2. Next, polynomial equations including more coefficients (C_1, C_2, C_3, C_4, etc.) were used to find the best fit of the data points. This was continued until adding another higher-order term did not significantly increase R^2 (Fig. 2A).

The second step was to perform an ANOVA to select an equation from the generated set, which could predict the EPWS values significantly better. It was based not only on quality of fit but also on the physical meaning of the prediction equations (Motulsky & Christopoulos, 2004). High values of R^2 associated with a P-value < 0.01 indicated the third-order polynomial as the best-fitting equation that could provide the best approximation to the experimental values of EPWS.

In the third step it was evaluated how the selected best-fitting, in this case the polynomial equation fits the EPWS data significantly better than a linear equation. Linear and third-order polynomial predictions of the EPWS were superimposed on experimental values together with their respective SD (Fig. 2B). The quality of these regressions was assessed by examining the respective residual plots. The linear equation was inappropriate for the description of the EPWS data because residuals clustering indicated that the data differed systematically (not just randomly) from the prediction curves. Positive residuals tended to cluster together at the first thoracic and the last lumbar vertebrae, whereas negative residuals clustered together in the transition zone from the thoracic to the lumbar region. In contrast, polynomial residual plots had a random

arrangement of residuals, which was more appropriate to predict EPWS (Fig 2B).

The fourth step corresponds to the evaluation of the predictability of the best-fitting equation of the set of equations developed in the third step using experimental data from two further datasets. The dataset of Berry et al. (1987) includes 12 vertebral parameters of three thoracic vertebrae (T2, T7 and T12) and four lumbar vertebrae (L1–4). The dataset of Scoles et al. (1988) includes 10 vertebral parameters of five thoracic vertebrae (T1, T3, T6, T9 and T12) and two lumbar vertebrae (L1 and L3) of male and female data.

Results

In general, there were no large differences for the correlations of each of the individual 20 vertebral parameters with VBHP when comparing linear, exponential and logarithmic regressions with each other. For this reason, only the linear predictions are provided from these results (Figs 3, 4 and 5). High correlations were found for parameters related to endplates (EPWS, EPWI, EPDS and EPDI) and vertebral bodies (VBW and VBD) (R^2, from 0.923 to 0.959) (Fig. 3) (please see Fig. 1 for all abbreviations). Moderate values of R^2 (from 0.520 to 0.793) were achieved in pedicle heights (PHL and PHR) and transverse inclination left (PTIL) (Fig. 4)

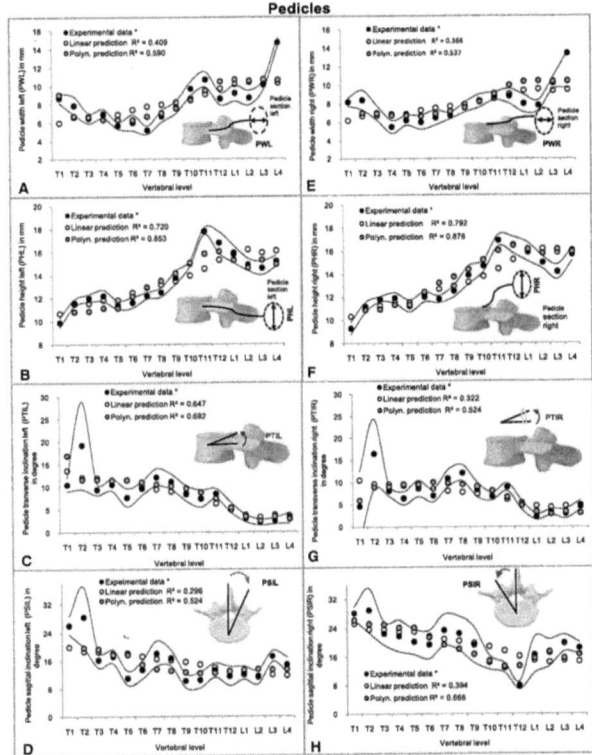

Fig. 4 Linear and polynomial predictions of parameters related to pedicles (PWL, PWR, PHL, PHR, PTIL, PTIR, PSIL and PSIR) (A-H) superimposed on experimental data of Panjabi et al. (1991, 1992)*. Dotted curve indicates SD of the experimental data.

as well as in the spinal canal (SCW and SCD) (Fig. 5). However, about 50% of the investigated parameters showed low or no correlation with VBHP. These were the dimensions of endplate inclinations (EPIS and EPII) (Fig. 3), pedicles (PWL, PWR, PTIR, PSIL and PSIR) (Fig. 4) and other posterior structures (SPL and TPW) (Fig. 5).

In contrast to the above regressions, third-order polynomial regressions provided the best results with significant correlations between all selected parameters and VBHP (Table 1). As the dataset of Panjabi et al. (1991, 1992) does not include VBW and VBD, an alternative method for the prediction of these parameters was used. The inclusion of more than four coefficients increased the R^2 values but ANOVAs indicated that the obtained correlations did not significantly improve parameter predictions. For instance, fourth- and fifth-order polynomial regressions between the PWL and VBHP resulted in P-values > 0.05.

The parameters EPWS, EPWI, EPDS, EPDI, VBW and VBD that showed high correlations by linear, logarithmic and exponential regressions exhibited, after third-order polynomial regressions, an increase of R^2 (from 0.970 to 0.982, P-values < 0.01) (Fig. 3). Similarly, the correlations with PHL, PHR, PTIL, SCW and SCD were improved and R^2 values ranging from 0.693 to 0.964 were achieved (Figs 4 and 5). Furthermore, for those parameters that displayed low or no correlation with anterior procedures (EPIS, EPII, PWL, PWR, PTIR, PSIL, PSIR, SPL and TPW), polynomial regressions achieved reasonable correlations (R^2, from 0.514 to 0.693, P-values < 0.05) (Figs 3, 4 and 5). An exception was PTIL, for which the best results were obtained after exponential regression (R^2 = 0.757) (Fig. 4C).

The prediction of the vertebral parameters related to anterior vertebral structures using linear, exponential, logarithmic and polynomial prediction equations did not demonstrate significant differences (Fig. 6). Moreover, polynomial prediction equations were required to predict the parameters related to posterior vertebral structures. The polynomial predictions are generally within or close to

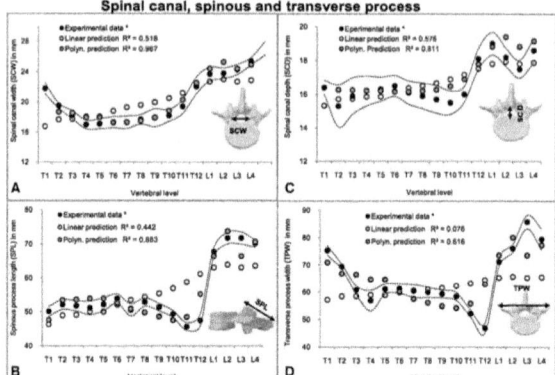

Fig. 5 Linear and polynomial predictions of other vertebral posterior structures (SCW, SCD, SPL and TPW) (A-D) superimposed on experimental data of Panjabi et al. (1991, 1992)*. Dotted curve indicates SD of the experimental data.

the regions of the 95% confidence intervals of the experimental data of Panjabi et al. (1991, 1992).

Using the dataset of Berry et al. (1987), a comparison of predicted EPWS and EPDI showed mean percent errors of −14.93 and 31.86%, respectively for T1; all other levels were very close to experimental data with mean percent errors of −0.32 to 9.68% (Fig. 7A). Predictions of PHL showed better results for thoracic levels with the smallest error being −0.05 mm (−0.8%) for PHL (T2) and a mean percent error of approximately 22.5% for thoracic and 24.6% for lumbar

Table 1 Polynomial coefficients (C_1, C_2, C_3 and C_4) for prediction equations of 20 parameters per vertebral level of the human thoracic and lumbar spine.

Vertebral parameter		Abbreviation	C_1	C_2	C_3	C_4	SD	R^2	P-value
Endplate	Width	EPWS	121.650	−15.403	0.742	−0.010	1.195	0.982	1.07E−10
		EPWI	300.140	−43.509	2.206	−0.035	1.454	0.976	5.19E−10
	Depth	EPDS	−60.076	8.983	−0.293	0.004	0.852	0.981	1.31E−10
		EPDI	−63.590	9.473	−0.300	0.003	0.769	0.981	1.26E−10
	Inclination	EPIS	−66.833	12.035	−0.691	0.013	0.699	0.606	0.008981
		EPII	66.233	−9.095	0.418	−0.006	0.606	0.514	0.030367
Vertebral body	Width	VBW*	4.1409	0.748	–	–	1.287	0.969	5.69E−05
	Depth	VBD**	−80.223	10.313	−0.350	0.004	0.523	0.995	0.000666
Pedicle	Width	PWL	230.261	−34.915	1.777	−0.029	1.682	0.590	0.011194
		PWR	157.740	−23.284	1.168	−0.019	1.446	0.537	0.022469
	Height	PHL	168.200	−27.194	1.522	−0.027	0.954	0.853	2.80E−05
		PHR	105.820	−17.256	0.999	−0.018	0.872	0.879	8.65E−06
	Transverse inclination	PTIL	−10.658	2.9889	−0.099	−0.001	2.741	0.693	0.002089
		PTIR	−202.510	31.347	−1.496	0.023	2.877	0.524	0.026147
	Sagittal inclination	PSIL	305.290	−39.194	1.734	−0.025	4.064	0.524	0.026141
		PSIR	−275.130	53.937	−3.119	0.057	3.403	0.669	0.003290
Spinal canal	Width	SCW	206.750	−26.838	1.218	−0.017	0.634	0.964	6.76E−09
	Depth	SCD	−2.449	3.8232	−0.254	0.006	0.573	0.811	0.000123
Spinous process	Length	SPL	−947.110	168.10	−9.310	0.170	3.472	0.882	7.43E−06
Transverse process	Width	TPW	−343.670	80.885	−5.090	0.102	7.259	0.616	0.007667

SD in mm (for linear dimensions) or in degree (for angular dimensions).
The basic form of the prediction equations is $y = C_1 + C_2x + C_3x^2 + C_4x^3$ where y is the value of the parameter to be predicted and x is the value of the VBHP on each vertebral level.
S, superior; I, inferior; L, left; R, right.
*For VBW, x is the value of EPWS/EPWI on each vertebral level.
**For VBD, x is the value of EPDs/EPDI on each vertebral level.

© 2009 The Authors
Journal compilation © 2009 Anatomical Society of Great Britain and Ireland

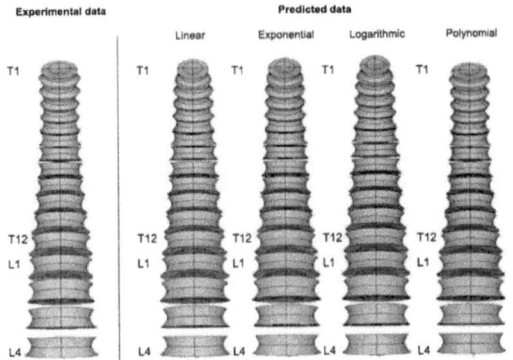

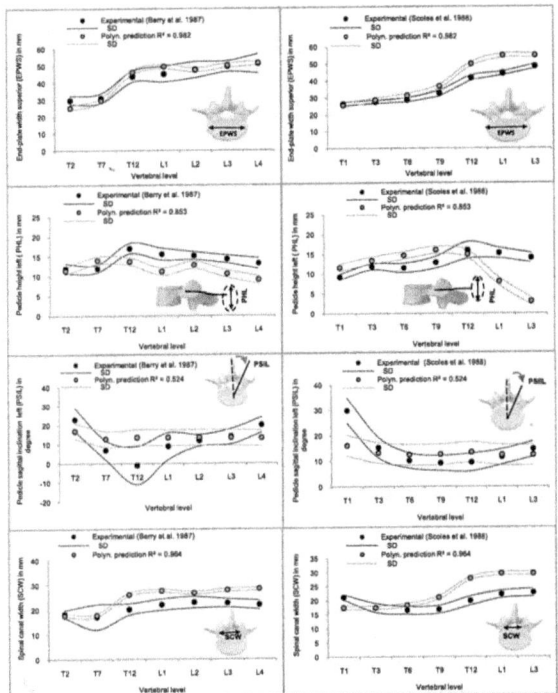

Fig. 6 Geometric models of the human thoracic (T1–12) and lumbar (L1–4) vertebrae constructed with parameters related to endplates and vertebral bodies (EPWS, EPWI, EPDS, EPDI, EPIS, EPII, VBW and VBD). The first model corresponds to the data of Panjabi et al. (1991, 1992) and was created using eight parameters per vertebral level (a total of 128 parameters). The other models were generated using only the values of the VBHP of each vertebral level and predicted parameters from linear, exponential, logarithmic and polynomial equations.

Fig. 7 Comparison of some predicted vertebral parameters (EPWS, PHL, PSIL and SCW) with corresponding experimental data from Berry et al. (1987) (left column, A–D) and Scoles et al. (1988) (right column, E–H) in selected vertebral levels. The means and 95% confidence intervals (dotted lines) of the experimental and predicted values are shown.

levels (Fig. 7B). Polynomial pedicle predictions showed a high error for PSIL (T12) (Fig. 7C). Predictions related to the SCW and SCD also displayed better results for thoracic levels with the largest error being 1.19 mm (7.93%). Lumbar levels showed an approximate mean percent error of 23% (SCW) and 31% (SCD) (Fig. 7D).

With the dataset of Scoles et al. (1988), predictions of EPWS and VBD showed a range of mean percent errors of −2.68 to 23.48%, with the largest errors occurring in EPWS (L1) (Fig. 7E). The shortest error for PHL was −0.69 mm (−4.66%) for T12 and a mean percent error of approximately 18.5% for thoracic and 45.6% for lumbar levels was found (Fig. 7F). Polynomial pedicle predictions showed a high error for PSIL (T1) (Fig. 7G). Predictions of SCW and SCD showed similar results to the prediction with the dataset of Berry et al. (1987), with the largest error being 4.25 mm (22.1%) for thoracic levels (Fig. 7H).

Discussion

Linear and nonlinear regression analyses were performed with the anatomical data of Panjabi et al. (1991, 1992) to generate prediction equations for 20 vertebral parameters per vertebral level of the human thoracic (T1–12) and lumbar (L1–4) vertebrae as a function of the VBHP. The parameters corresponding to the vertebra L5 were not included in the analyses because L5 shows remarkable morphological differences for some parameters when compared with the other lumbar vertebrae (Berry et al. 1987; Zindrick et al. 1987; Scoles et al. 1988). This is probably due to the position of L5 being localized in the final transition zone, from lumbar to sacral region (Panjabi et al. 1989, 1992).

In this study two assumptions were necessary. First, despite the high anatomical variability of the human vertebrae, the dataset of Panjabi et al. (1991, 1992) was assumed to be representative of the adult population without spinal pathology. Second, it was assumed that the dimensions of the vertebral structures described in this dataset were obtained precisely. As the three datasets used in this study were provided from *in-vitro* measurements, further investigations are necessary to evaluate the predictability of the regression equations with a dataset from patients.

Third-order polynomial equations represented the best regression approximation as indicated after analysis of covariance (Table 1). SEs indicated that, with few exceptions, such as for pedicle inclinations, the best fit values for the prediction equations were accomplished with reasonable certainty. Pedicle inclinations showed a wide variation that can be observed in the wide confidence interval of the sagittal plane angle for the mid-thoracic vertebrae (Fig. 4C,D,G,H).

Our results were compared, when possible, with existing published data. All correlation coefficients generated using polynomial regressions were considerably better than the values obtained by Breglia (2006) using simple linear regressions on the data of Panjabi et al. (1991, 1992). The parameters related to posterior structures that could not be predicted with the regressions of Breglia (2006) have shown a moderate correlation after polynomial regressions. Linear regression procedures are straightforward and the results appear to be readily evaluated statistically. However, the relationships between the vertebral variables follow a curved line, not a straight line. Although the methods used for fitting a nonlinear equation such as polynomial regression are extensions of linear regression, the results are better because polynomial equations can be used to create a generic curve through the data points; more coefficients create a more flexible curve, which could better fit the data.

Comparisons of the theoretical predictions with two other sets of experimental data (Berry et al. 1987; Scoles et al. 1988) indicated that the predictions generally agree well with the experimental data. Although the differences in the predictions of pedicle inclination (Fig. 7C,G) have been relatively great, a reasonable correlation between the main posterior elements and VBHP was found. This is not in accordance with Scoles et al. (1988) who declared that it was not possible to establish useful predictors for pedicle dimensions based on the size of the vertebral body. Differences in predicted values may also be attributed to technical factors related to obtaining these anatomical data, such as different protocols of preparation and measurement. Furthermore, there are individual variations and aging that can induce substantial changes in each individual's vertebrae (Bernick & Cailliet, 1982; Diacinti et al. 1995).

Lavaste et al. (1992) developed a method to reconstruct lumbar vertebral geometry from two X-rays (anterior–posterior and lateral) using multiple-linear regression analysis. However, to predict the vertebral geometry, six given parameters per vertebra were required. A digitalization process to define these parameters showed a relative error of approximately 15%. Moreover, the orientation and width of the pedicles were not taken into consideration. Laporte et al. (2000) performed a similar study in thoracic vertebrae, which required the measurement of 15 parameters per vertebra by X-ray in order to explain 100% of the variability for each parameter.

The advantage of using the generated set of prediction equations (Table 1) is the capability to model vertebral geometry in each level of the thoracic and lumbar spine, with the exception of L5, using only one parameter per vertebrae (VBHP), which can be easily measured on conventional lateral X-rays.

Conclusion

The present study shows that nonlinear regression analyses provide a time-efficient approach for modeling of the human vertebrae, allowing a better understanding of statistical correlations between vertebral dimensions. The geom-

© 2009 The Authors
Journal compilation © 2009 Anatomical Society of Great Britain and Ireland

etry that was reconstructed using the predicted vertebral parameters may be applied for the construction of finite element models of the spine without the need for expensive, invasive and time-consuming data collection, such as medical images. Another advantage is that this approach allows the values of the vertebral parameters to be changed, producing different vertebral morphologies. This could be used for the development of parameterized models of the spine to perform studies based on geometry variation, such as in spinal deformations, including scoliosis.

Acknowledgements

This study was financially supported by the German Research Foundation (Wi-1352/12-1).

Conflict of interest statement

Each author of this study did not and will not receive benefits in any form from a commercial party related directly or indirectly to the content of this study.

References

Aharinejad S, Bertagnoli R, Wicke K, et al. (1990) Morphometric analysis of vertebrae and intervertebral discs as a basis of disc replacement. *Am J Anat* 189, 69–76.

Aubin C-É, Dansereau J, Parent F, et al. (1997) Morphometric evaluation of personalised 3D reconstructions and geometrical models of the human spine. *Med Biol Eng Comput* 35, 611–618.

Bernick S, Cailliet R (1982) Vertebral endplate changes with aging of human vertebrae. *Spine* 7, 92–97.

Berry JL, Moran JM, Berg WS, et al. (1987) A morphometric study of human lumbar and selected thoracic vertebrae. *Spine* 12, 362–367.

Breglia DP (2006) *Generation of a 3-D Parametric Solid Model of the Human Spine Using Anthropomorphic Parameters*. Master dissertation. Ohio: Ohio University.

Cotterill PC, Kostuik JP, D'Angelo GD, et al. (1986) An anatomical comparison of the human and bovine thoracolumbar spine. *J Orthop Res* 4, 298–303.

Diacinti D, Acca M, D'Erasmo E, et al. (1995) Aging changes in vertebral morphometry. *Calcif Tissue Int* 57, 426–429.

Dumas R, Le Bras A, Champain N, et al. (2004) Validation of the relative 3D orientation of vertebrae reconstructed by bi-planar radiography. *Med Eng Phys* 26, 415–422.

Fagan MR, Julian S, Mohsen AM (2002) Finite element analysis in spine research. *Proc Inst Mech Eng H* 216, 281–298.

Gilbertson LG, Goel VK, Kong WZ (1995) Finite element methods in spine biomechanics research. *Crit Rev Biomed Eng* 23, 411–473.

Hall LT, Esses SI, Noble PC (1998) Morphology of the lumbar vertebral endplates. *Spine* 23, 1517–1522.

Krag MH, Weaver DL, Beynnon BD, et al. (1988) Morphometry of the thoracic and lumbar spine related to transpedicular screw placement for surgical spinal fixation. *Spine* 13, 27–32.

Laporte S, Mitton D, Ismael B, et al. (2000) Quantitative morphometric study of thoracic spine. A preliminary parameters statistical analysis. *Eur J Orthop Surg Traumatol* 10, 85–91.

Lavaste F, Skalli W, Robin S, et al. (1992) Three-dimensional geometrical and mechanical modelling of the lumbar spine. *J Biomech* 25, 1153–1164.

Marchesi D, Schneider E, Glauser P, et al. (1988) Morphometric analysis of the thoracolumbar and lumbar pedicles, anatomoradiologic study. *Surg Radiol Anat* 10, 317–322.

Masharawi Y, Rothschild B, Dar G, et al. (2004) Facet orientation in the thoracolumbar spine. *Spine* 29, 1755–1763.

Moran JM, Berg WS, Berry JL, et al. (1989) Transpedicular screw fixation. *J Orthop Res* 7, 107–114.

Motulsky H, Christopoulos A (2004) *Fitting Models to Biological Data Using Linear and Nonlinear Regression: A Practical Guide to Curve Fitting*, pp. 32–37. New York: Oxford University Press.

Panjabi MM, Yamamoto I, Oxland T, et al. (1989) How does posture affect coupling in the lumbar spine? *Spine* 14, 1002–1011.

Panjabi MM, Takata K, Goel V, et al. (1991) Thoracic human vertebrae – Quantitative three-dimensional anatomy. *Spine* 16, 889–901.

Panjabi MM, Goel V, Oxland T, et al. (1992) Human lumbar vertebrae – Quantitative three-dimensional anatomy. *Spine* 17, 299–306.

Pomero V, Mitton D, Laporte S, et al. (2004) Fast accurate stereographic 3D-reconstruction of the spine using a combined geometric and statistic model. *Clin Biomech* 19, 240–247.

Robin S, Skalli W, Lavaste F (1994) Influence of geometrical factors on the behavior of lumbar spine segments: a finite element analysis. *Eur Spine J* 3, 84–90.

Scoles PV, Linton AE, Buce L, et al. (1988) Vertebral body and posterior element morphology: the normal spine in middle life. *Spine* 13, 1082–1086.

Semaan I, Skalli W, Veron S, et al. (2001) Quantitative 3D anatomy of the lumbar spine. *Rev Chir Orthop Reparatrice Appar Mot* 87, 340–353.

Zindrick MR, Wiltse LL, Doornik A, et al. (1987) Analysis of the morphometric characteristics of the thoracic and lumbar pedicles. *Spine* 12, 162–166.

Prediction of the human thoracic and lumbar articular facet joint morphometry from radiographic images

Maria E. Kunkel, Hendrik Schmidt and Hans-Joachim Wilke

Institute of Orthopaedic Research and Biomechanics, University of Ulm, Ulm, Germany

Abstract

The articular facet joints (AFJ) play an important role in the biomechanics of the spine. Although it is well known that some AFJ dimensions (e.g. facet height/width or facet angles) play a major role in spinal deformities such as scoliosis, little is known about statistical correlations between these dimensions and the size of the vertebral bodies. Such relations could allow patient-specific prediction of AFJ morphometry from a few dimensions measurable by X-ray. This would be of clinical interest and could also provide parameters for mathematical modeling of the spine. Our purpose in this study was to generate prediction equations for 20 parameters of the human thoracic and lumbar AFJ from T1 to L4 as a function of only one given parameter, the vertebral body height posterior (VBHP). Linear and nonlinear regression analyses were performed with published anatomical data, including linear and angular dimensions of the AFJ and vertebral body heights, to find the best functions to describe the correlations between these parameters. Third-order polynomial regressions, in contrast to the linear, exponential and logarithmic regressions, provided moderate to high correlations between the AFJ parameters and vertebral body heights; e.g. facet height superior and interfacet width (R^2, 0.605–0.880); facet height inferior, interfacet height and sagittal/transverse angle superior (R^2, 0.875–0.973). Different correlations were found for facet width and transverse angle inferior in the thoracic (R^2, 0.703–0.930) and lumbar (R^2, 0.457–0.892) regions. A set of 20 prediction equations for AFJ parameters was generated (P-values < 0.005, ANOVA). Comparison of the AFJ predictions with experimental data indicated mean percent errors < 13%, with the exception of the thoracolumbar junction (T12–L1). It was possible to establish useful predictions for human thoracic and lumbar AFJ dimensions based on the size of the vertebral bodies. The generated set of equations allows the prediction of 20 AFJ parameters per vertebral level from the measurement of the parameter VBHP, which is easily performed on lateral X-rays. As the vertebral body height is unique for each person and vertebral level, the predicted AFJ parameters are also specific to an individual. This approach could be used for parameterized patient-specific modeling of the spine to explore the clinically important mechanical roles of the articular facets in pathological conditions, such as scoliosis.

Key words articular facet joints; human anatomy; spinal morphology; zygapophyseal joints.

Introduction

The articular facet joints (AFJ) play an important role in the biomechanics of the spine; they transmit a significant percentage of spinal loading, and provide translational, rotational and axial stability in the spine (Lorenz et al. 1983; Adams & Hutton, 1983; Onan et al. 1998). The shape, position and orientation of the AFJ strongly regulate the physiological motion of the spine (Taylor & Twomey, 1986; White & Panjabi, 1990). A patient-specific prediction of vertebral morphometric data (e.g. size and orientation of the AFJ) could be of clinical interest in the evaluation of operative and non-operative spinal treatments. For example, during clinical or radiographic examination of a patient with a spinal deformity such as scoliosis, the physicians are faced with the problem that it is not possible to know what the original anatomical dimensions of this spine were before deformity started. Most procedures for spinal deformity corrections are based on the average values of vertebral dimensions of a healthy spine, without taking into account that each patient has a specific morphometry. One possibility would be to use an X-ray of a patient and, from direct measurement of a dimension of an intact vertebra (e.g. vertebral body height), predict other dimensions (e.g.

Correspondence

Hans-Joachim Wilke, Institute of Orthopaedic Research and Biomechanics, Helmholtzstraße 14, D-89081 Ulm, Germany.
T: + 49 731 50055320; F: + 49 731 50055302;
E: hans-joachim.wilke@uni-ulm.de

Accepted for publication 22 October 2010

© 2010 The Authors
Journal of Anatomy © 2010 Anatomical Society of Great Britain and Ireland

articular facet size or orientation) by using the relationships between them.

Another important issue for such predictions of vertebral morphometry is to provide parameters for mathematical modeling of the spine. Shirazi-Adl (1991, 1994), Zander et al. (2003) and Schmidt et al. (2008a,b, 2009) investigated the AFJ in normal and pathological conditions using finite element models. The geometry of these models often comes from 3D reconstruction of computed tomography, which requires a considerable processing time and computational power. Finite element models with a simple but parameterized geometry could be an alternative to these complex models because they would allow different vertebral geometries to be fitted.

The human vertebral morphometry has been measured in a number of studies (Table 1) and using patient-specific stereoradiographic reconstruction techniques (Aubin et al. 1997; Petit et al. 1998; Pomero et al. 2004). Previous investigations of statistical correlations between dimensions of vertebral structures have been used to predict vertebral dimensions. However, the relationships between the geometrical parameters describing the different parts of a given vertebra and the vertebral body height are only well known for the anterior structures and some posterior structures (e.g. pedicles) (Scoles et al. 1988; Kunkel et al. 2010).

Although it is well known that some AFJ parameters, such as facet height and width or facet angles, play a major role in spinal deformities such as scoliosis due to pathological changes in the anatomy of the posterior structures (Parent et al. 2002; Aebi, 2005), little is known about vertebral relationships between AFJ parameters and vertebral heights. Breglia (2006) investigated only two AFJ parameters and found a poor correlation between them and the vertebral body height posterior (VBHP). Other studies investigated statistical correlations of AFJ parameters not only with VBHP but with all other vertebral parameters; in these cases, 6–15 initial parameters were necessary for the prediction of some AFJ parameters (Lavaste et al. 1992; Skalli et al. 1993; Maurel et al. 1997; Laporte et al. 2000). A quantitative analysis of the relationship between the AFJ parameters and VBHP allowing patient-specific predictions has not yet been investigated.

The purpose of the current study was to generate prediction equations for 20 parameters of the human thoracic

Table 1 Summary of *in vitro* measurements of geometric parameters of the human articular facets reported for thoracic and lumbar vertebrae.

Reference	Technique	Vertebral level	Facet geometric parameters
van Schaik et al. (1985)	CT	L3–5	Transverse plane angle
Cotterill et al. (1986)	CT	T6, T12, L3	Interfacet height/width
			Transverse plane angle
Berry et al. (1987)	DM	T2, T7, T12, L1–5	Interfacet height
Scoles et al. (1988)	DM	T1, T3, T6, T9, L1, L3, L5	Inferior facet pedicle distance
			Inferior facet/mid-pedicle length
Ahmed et al. (1990)	CT	L2–5	Facet angle/depth/length
Panjabi et al. (1993)		T1–12, L1–5	Facet height/width
			Interfacet height/width
			Facet area
			Transverse/sagittal plane angle
Boszczyk (1997)	DM	T12, L1–5	Facet height/width
			Transverse/Sagittal plane angle
Ebraheim et al. (1997)	DM	T1–12	Facet thickness
			Facet height/width
			Sagittal plane angle
Laporte et al. (2000)	DM	T1–12	Transverse/Sagittal plane angle
			Facet height/width
			Facet area
			Distance from facet centers
Dai (2001)	CT/MRI	L4–5	Transverse/Sagittal plane angle
Masharawi et al. (2004)	DM	T1–12, L1–5	Transverse/sagittal plane angle
			Interfacet height/width
Masharawi et al. (2005)	DM	T1–12, L1–5	Facet height/width
Masharawi et al. (2007a)		L1–5	Facet height/width
			Interfacet height/width
			Facet concavity/convexity
Masharawi et al. (2007b)	DM	L1–5	Transverse plane angle
Wang & Yang (2009)	CT	L4–5	Sagittal plane angle

CT, computed tomography; DM, direct measurement; MRI, magnetic resonance imaging.

and lumbar AFJ as a function of only one given parameter measurable by X-ray, the VBHP.

Materials and methods

Study population

Vertebral anatomical data were collected from Panjabi et al. (1991, 1992, 1993) and were included in this study. To carry out statistical analyses, measurements of the VBHP and 20 anatomical measurements were selected to describe the size and orientation of the human thoracic (T1–12) and lumbar (L1–4) AFJ (Fig. 1).

Statistical analysis

To perform linear and nonlinear regression with these data, the same methodology used by Kunkel et al. (2010) was adopted. The parameter VBHP was used as the predictor variable, and the values of the parameters related to the vertebral level L5 were not included in the analysis. However, due to the existence of a considerable difference between the morphometry of the articular facets in the thoracic and lumbar regions, a modification was introduced to achieve a better prediction of the AFJ parameters. It refers to separate regressions using thoracic (T1–12) and lumbar (L1–4) data together, thoracic data alone, and lumbar data alone. A brief description of the implemented methodology in this study follows.

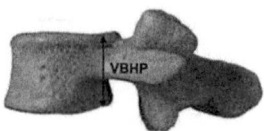

VBHP   vertebral body height posterior

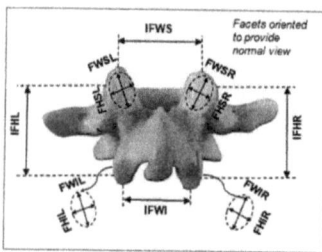

Facet linear parameters

FHSL	facet height superior left
FHSR	facet height superior right
FHIL	facet height inferior left
FHIR	facet height inferior right
FWSL	facet width superior left
FWSR	facet width superior right
FWIL	facet width inferior left
FWIR	facet width inferior right
IFHL	interfacet height left
IFHR	interfacet height right
IFWS	interfacet width superior
IFWI	interfacet width inferior

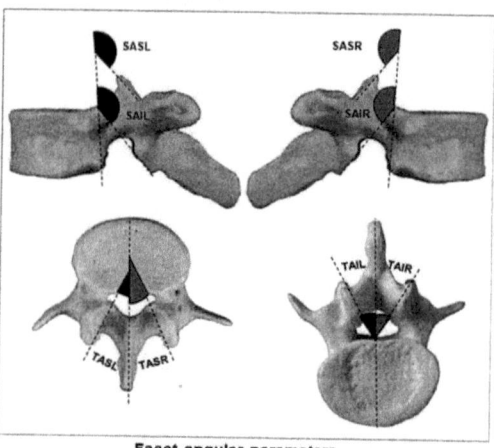

Facet angular parameters

SASL	sagittal angle superior left
SASR	sagittal angle superior right
SAIL	sagittal angle inferior left
SAIR	sagittal angle inferior right
TASL	transverse angle superior left
TASR	transverse angle superior right
TAIL	transverse angle inferior left
TAIR	transverse angle inferior right

Fig. 1 Schematic representation of the anatomical parameters that were considered for linear and nonlinear regression analyses. Vertebral body height posterior and linear and angular dimensions of the articular facet joints. The facet surfaces were approximated by a plane. The orientation of the planes was defined by two angles made by the facet plane with the sagittal and transverse anatomic planes.

Correlation and regression analyses with AFJ parameters and VBHP

Each AFJ parameter selected for this study from T1–12 and L1–4 was correlated with VBHP. A least-squares regression analysis was performed to find linear and nonlinear functions to describe the relationship between each pair of parameters. These functions represented prediction equations which take the forms $y = C_1 + C_2 x$ for linear, $y = C_1 + C_2 ln(x)$ for logarithmic,

$$y = C_1 e^{C_2 x}$$

for exponential, and

$$y = C_1 + C_2 x + C_3 x^2 + C_4 x^3 + \ldots C_n x^{n-1}$$

for polynomial equations, where y was the value of the articular facet parameter to be predicted, x was the value of the VBHP on each vertebral level, and $C_1, C_2, C_3, C_4 \ldots C_n$ were the regression coefficients. The fraction of the overall variance of each facet parameter that was reduced by a specific regression, R^2 value, was used to assess the best function of the set of generated prediction equations. For the polynomial equations, the number of coefficients was continually increased until adding another higher-order term did not significantly increase R^2.

Identification of the best prediction equation

An ANOVA was performed to define the significance of each prediction equation that fits a given facet parameter. The respective residual plots were examined to assess the quality of these equations. The identification of the best prediction equations was based on the following criteria: values of $R^2 > 0.5$ associated with a probability level of P-value < 0.05 and a standard error of the estimate (SE) < 30% of the standard deviations of the experimental data of Panjabi et al. (1993).

Evaluation of the predictability of the prediction equation

For each AFJ parameter, the theoretical predictions were plotted against the corresponding known experimental data of Panjabi et al. (1993) and Cotterill et al. (1986) considering their respective confidence limits. The dataset of Cotterill et al. (1986) included three AFJ parameters (interfacet height and width, and transverse angle superior) of two thoracic (T6 and T12) and one lumbar (L3) vertebrae. As these three parameters provided in this dataset did specify where the measurements were obtained (right or left side, superior or inferior), the predictions of these parameters in our study were performed considering all possibilities.

Results

Third-order polynomial regressions, in contrast to the linear, exponential, logarithmic and polynomial regressions with other orders, provided the best results with significant

Table 2 Polynomial coefficients (C_1, C_2, C_3 and C_4) for prediction equations of 20 articular facet joint parameters per vertebral level of the human thoracic spine. Refer to Fig. 1 for abbreviations.

Parameter		Abbreviation	C_1	C_2	C_3	C_4	SE	R^2	P-value
Facet dimensions									
	Height	FHSL	−4.775	4.189	−0.309	0.007	1.18	0.647	0.0046
		FHSR	33.916	−2.520	0.068	0.001	0.87	0.751	0.0006
		FHIL	62.954	−6.484	0.235	−0.002	0.68	0.910	1.49E-06
		FHIR	82.879	−10.102	0.444	−0.006	0.64	0.924	5.61E-07
	Width	FWSL	381.760	−59.777	3.175	−0.056	0.7	0.771	0.0061
		FWSR	244.770	−36.675	1.890	−0.032	0.46	0.850	0.0011
		FWIL	383.000	−61.279	3.325	−0.059	0.60	0.788	0.0045
		FWIR	314.610	−50.508	2.763	−0.050	0.60	0.703	0.0164
	Interfacet height	IFHL	45.467	−5.543	0.329	−0.005	0.79	0.970	2E-06
		IFHR	−56.782	11.697	−0.628	0.013	0.76	0.973	1.35E-06
	Interfacet width	IFWS	744.410	−111.100	5.625	−0.094	1.47	0.880	0.0005
		IFWI	620.110	−95.960	5.0701	−0.088	0.89	0.868	0.0007
Facet orientations									
	Transverse angle	TASL	−629.850	106.150	−5.322	0.089	2.09	0.922	5.71E-07
		TASR	−468.340	81.873	−4.122	0.070	1.70	0.938	1.65E-07
		TAIL	−772.290	131.960	−6.800	0.116	1.40	0.930	5.73E-05
		TAIR	−294.800	53.407	−2.529	0.040	1.98	0.785	0.0047
	Sagittal angle	SASL	998.880	−149.700	8.0381	−0.143	1.04	0.952	1.31E-05
		SASR	−1084.000	160.740	−8.4545	0.147	1.41	0.904	0.0002
		SAIL	−2549.600	463.78	−27.063	0.521	8.47	0.849	0.0012
		SAIR	2375.500	−434.840	25.490	−0.493	8.20	0.862	0.0084

SE in mm (for facet linear parameter) or in degree (for facet angular parameters).
The basic form of the prediction equations is $y = C_1 + C_2 x + C_3 x^2 + C_4 x^3$, where y is the value of the facet parameter to be predicted and x is the value of the VBHP on each vertebral level. S, superior; I, inferior; L, left; R, right. Values in bold indicate facet parameters that show the same polynomial coefficients for thoracic and lumbar vertebrae.

correlations between each of the AFJ parameters and VBHP. The inclusion of more than four polynomial coefficients increased the R^2 values; however, ANOVA S indicated that the obtained correlations did not significantly improve parameter predictions. For this reason, only the results of third-order polynomial predictions are provided (Table 2 for thoracic and Table 3 for lumbar regions).

Considering the previously established criteria, the polynomial regressions using the thoracic and lumbar data together (T1–12 and L1–4) showed variable correlations with VBHP (R^2 0.516–0.950), providing significant prediction equations for all selected AFJ parameters from VBHP (an exception was FWSL).

The polynomial regressions considering only the thoracic data (T1–12) resulted in improvements for predictions of 70% of the thoracic AFJ parameters such as an increase of R^2 (from 0.650 to 0.973), an increase in the significance and a decrease of SE (Table 2). In the case of the other thoracic AFJ parameters, such as facet heights (FHSL, FHSR, FHIL and FHIR) and transverse plane angles (TASL and TASR), where no improvement was achieved, the polynomial coefficients obtained from the regressions using the thoracic and lumbar data together were accepted.

Polynomial regressions considering only lumbar data (L1–4) did not satisfy the minimum criteria required for the selection of the prediction equations. This was probably due to the low number of observations in the lumbar region (just four parameter values per vertebrae). Therefore, the polynomial coefficients obtained from the regressions using the thoracic and lumbar data together were used to predict the lumbar AFJ parameters (Table 3).

Facet linear parameters predictions

High correlations were found for inferior facet heights (FHIL and FHIR) with VBHP (R^2, 0.910–0.924). Superior facet heights (FHSL and FHSR) showed moderate correlations (R^2, 0.647–0.751). Comparisons with experimental data of Panjabi et al. (1993) showed that the best predictions were found for FHIR with a mean percent error of approximately 14% in T12 and a maximal error of 7% in all other levels (Fig. 2A). The largest error of approximately −17% was found for FHSL predictions in the lumbar level (Fig. 2D). For facet widths (FWSL, FWSR, FWIL, FWIR), no significant difference was found between the correlations of the widths of the superior and inferior facets with VBHP. The best correlations with VBHP were achieved on the thoracic levels (R^2, 0.703–0.850) where a mean percent error < 10% was found (e.g. Fig. 2B,E). Although these parameters

Table 3 Polynomial coefficients (C_1, C_2, C_3 and C_4) for prediction equations of 20 articular facet joint parameters per vertebral level of the human lumbar spine. Refer to Fig. 1 for abbreviations.

Parameter		Abbreviation	C_1	C_2	C_3	C_4	SE	R^2	P-value
Facet dimensions									
Height		FHSL	−4.775	4.188	−0.309	0.007	1.18	0.647	0.0046
		FHSR	33.916	−2.519	0.068	0.001	0.87	0.751	0.0006
		FHIL	62.954	−6.484	0.235	−0.002	0.68	0.910	1.49E-06
		FHIR	82.879	−10.102	0.444	−0.006	0.64	0.924	5.62E-07
Width		FWSL	240.180	−35.123	1.759	−0.029	1.49	0.457	0.0416
		FWSR	177.180	−25.063	1.233	−0.020	1.07	0.516	0.0283
		FWIL	109.750	−14.135	0.647	0.010	1.10	0.556	0.0255
		FWIR	70.833	−8.667	0.396	−0.006	1.09	0.559	0.0168
Interfacet height		IFHL	135.720	−20.994	1.201	−0.021	1.16	0.950	4.11E-08
		IFHR	97.381	−14.658	0.856	−0.015	1.33	0.942	1.06E-07
Interfacet width		IFWS	595.510	−85.587	4.185	−0.067	1.78	0.818	9.80E-05
		IFWI	312.070	−43.273	2.100	−0.033	2.50	0.605	0.0091
Facet orientations									
Transverse angle		TASL	−629.850	106.150	−5.322	0.089	2.09	0.922	5.71E-07
		TASR	−468.340	81.873	−4.122	0.070	1.70	0.938	1.65E-07
		TAIL	−578.400	98.978	−4.951	0.082	1.65	0.892	4.45E-06
		TAIR	−394.580	70.418	−3.485	0.057	3.24	0.526	0.0254
Sagittal angle		SASL	−1384.100	260.400	−15.230	0.292	10.44	0.875	1.02E-05
		SASR	895.720	−180.630	10.957	−0.217	10.08	0.876	9.99E-06
		SAIL	24.358	24.433	−2.355	0.064	14.45	0.808	0.0001
		SAIR	−88.384	−14.160	1.824	−0.055	13.93	0.829	6.74E-05

SE in mm (for facet linear parameter) or in degree (for facet angular parameters).
The basic form of the prediction equations is $y = C1 + C_2x + C_3x^2 + C_4x^3$, where y is the value of the facet parameter to be predicted and x is the value of the VBHP on each vertebral level. S, superior; I, inferior; L, left; R, right. Values in bold indicate facet parameters that show the same polynomial coefficients for thoracic and lumbar vertebrae.

© 2010 The Authors
Journal of Anatomy © 2010 Anatomical Society of Great Britain and Ireland

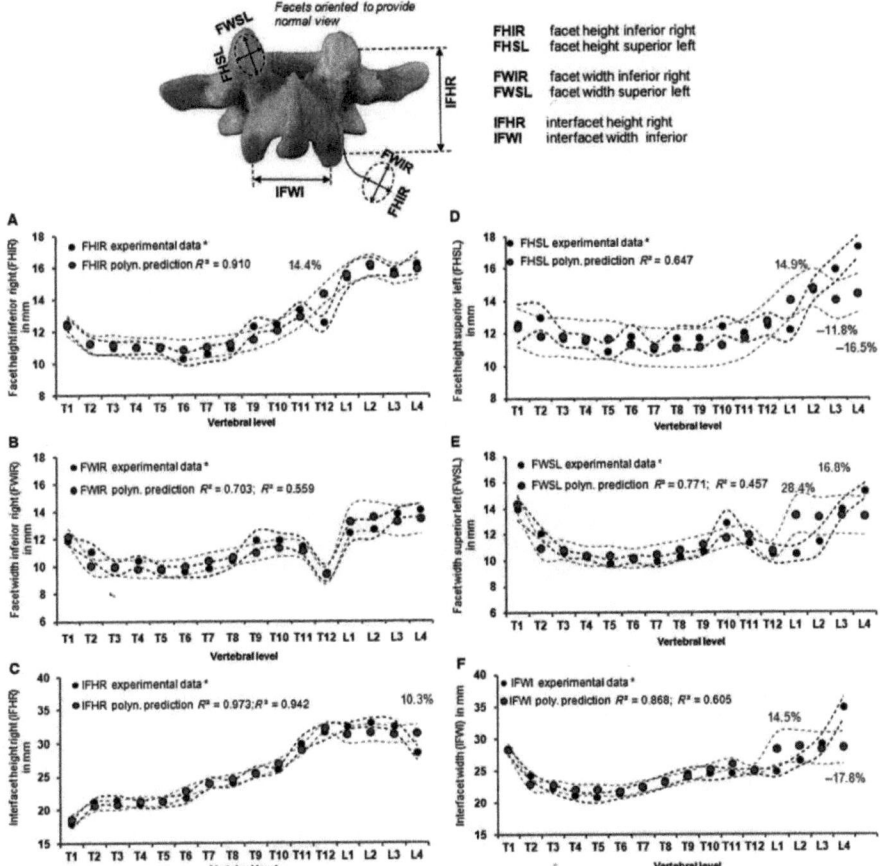

Fig. 2 Polynomial predictions of parameters related to dimensions of the human thoracic and lumbar articular facet joints (FHIR, FWIR, IFHR, FHSL, FWSL and IFWI) from VBHP, superimposed on experimental data of Panjabi et al. (1993)*. Dotted curves indicate standard deviation of both the predictions and experimental data. R^2 is provided for thoracic and for lumbar levels. Only mean percent errors > 5% for all vertebral levels are shown. Best predictions (left); worst predictions (right).

showed poor correlation for lumbar levels (R^2, 0.457–0.559) good predictions could be achieved with the largest error of 28.4% in the FWSL (L1) prediction (Fig. 2E). The interfacet heights and widths (IFHL, IFHR, IFWS and IFWI) exhibited high correlations with VBHP (R^2, 0.818–0.973) with the exception of thoracic IFWI (R^2 = 0.605). Independent of thoracic or lumbar levels, the best correlations were found for interface heights. Predictions of these parameters showed a mean percent error < 11% for all vertebral levels (e.g. Fig. 2C). The mean percent errors were < 12% (for thoracic) and < −18% (for lumbar) for the IFWS and IFWI (e.g. Fig. 2F).

Facet angular parameters predictions

The transverse angles showed high correlations for TASL, TASR and TAIL with VBHP for all levels (R^2, 0.785–0.938). TAIR displayed a poor correlation (R^2 = 0.526) for the lumbar levels. For the predictions with data of Panjabi et al. (1993) the largest mean percent error found was approximately 10% (TAIR, L4). All other predictions including lumbar levels showed an error < 6% (e.g. Fig. 3A,C). The sagittal angles superior and inferior (SASL, SASR, SAIL and SAIR) also showed high correlations with VBHP, with R^2 ranging from 0.808 to 0.952. For SASR and SASL, very good predictions using the data of Panjabi et al. (1993) were found for thoracic levels, with a mean percent error < 4%; the lumbar predictions for these parameters showed variable errors of −0.6 to −9.3% (e.g. Fig. 3B). SAIL and SAIR predictions in the T11 level showed a considerable mean percent error of approximately 27%; all other thoracic levels displayed an error < 9%. The lumbar predictions showed errors of 5–20% (e.g. Fig. 3D).

Evaluation of the prediction equations

The third-order polynomial predictions were generally within or close to the regions of the 95% confidence intervals of the experimental data of Panjabi et al. (1993) (Fig. 4). Using the dataset of Cotterill et al. (1986), the comparisons with experimental and predicted thoracic IFHR and IFHL showed a mean percent error of approximately 13% (T6) and an error of approximately 0.22 mm (< 2%) (T12) (Fig. 5A,D). The L3 prediction showed the largest error of approximately 2.6 mm (9%) for IFHL (Fig. 5D). The predictions of the IFWS and IFWI exhibited a mean percent error < 16% for the thoracic level and better results for L3, respectively 1.7 mm (6.3%) and 1.01 mm (3.8%) (Fig. 5B,E). The predictions of the facet orientation, TASR and TASL, showed very similar results to experimental data values, with mean percent errors of approximately −11% (T6), −5% (T12) and −9% (L3) (Fig. 5C,F).

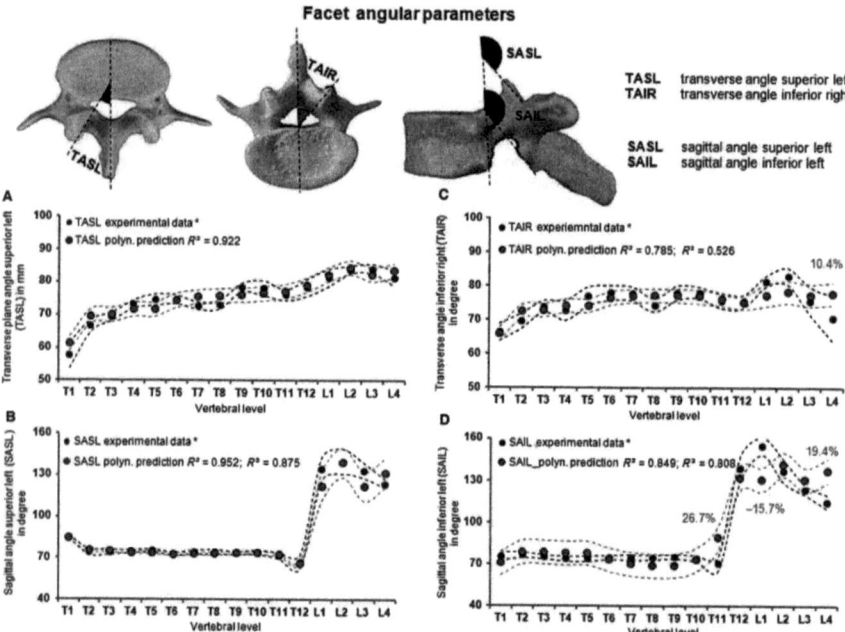

Fig. 3 Polynomial predictions of parameters related to orientations of the human thoracic and lumbar articular facet joints (TASL, SASL, TAIR and SAIL) from VBHP, superimposed on experimental data of Panjabi et al. (1993)*. Dotted curves indicate standard deviation of both the predictions and experimental data. R^2 is provided for thoracic and for lumbar levels. Only mean percent errors > 5% for all vertebral levels are shown. Best predictions (left); worst predictions (right).

© 2010 The Authors
Journal of Anatomy © 2010 Anatomical Society of Great Britain and Ireland

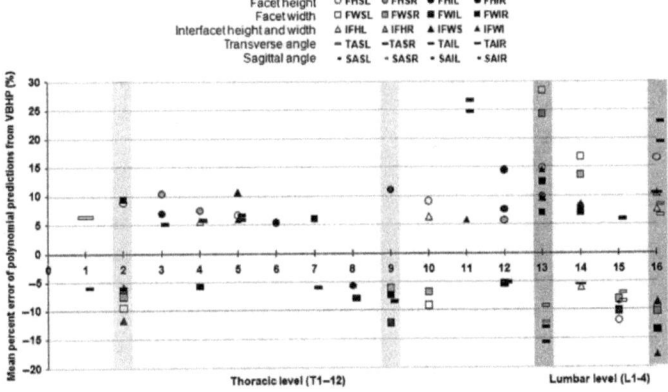

Fig. 4 Mean percent errors > 5% for polynomial predictions of 20 articular facet joint parameters from VBHP of experimental data of Panjabi et al. (1993). Error per vertebral level of the human thoracic (T1–12) and lumbar (L1–4) spine. Refer to Fig. 1 for abbreviations. Polynomial predictions were satisfactory for all AFJ parameters with a few exceptions of some thoracic facet linear dimensions (T2 and 9, light gray highlighted bars) and some lumbar facet orientations (L1 and 4, dark gray highlighted bars).

Discussion

In this study, the morphological relationships between anatomical characteristics of the human AFJ and VBHP were described by linear and nonlinear regression analyses. A set of prediction equations for AFJ parameters of the thoracic (T1–12) and lumbar (L1–4) spine was generated as a function of the VBHP (Tables 2 and 3).

SEs indicates that, with few exceptions, polynomial predictions were satisfactory for all AFJ parameters. The exceptions were in the facet dimensions in the thoracic region (T2 and 9) and facet orientation in the lumbar region (L1 and 4) (Fig. 4). The best predictions of facet orientations were found for the transverse angles. Excluding the level T1, no errors > 5% were found in the prediction of the transverse angles of the thoracic or lumbar regions, probably because these parameters show very little variability within the vertebrae from T1 to L4. Notably in the midthoracic region (i.e. T3–8), excellent predictions with errors < 10% could be achieved for most parameters of the AFJ. In contrast, predictions with errors of up to −15% were found for all sagittal angles in the thoracolumbar junction (T12–L1) (Fig. 4). This was due to the large variability of this region within individuals, with the AFJ being either frontally oriented, as in the thoracic vertebrae, or sagittally oriented, as in the lumbar vertebrae. This is in accordance with Goel & Weinstein (1990) and Masharawi et al. (2004) who showed that the morphology of the first lumbar vertebra is distinct from the other vertebrae, with a transition from the typically thoracic to the lumbar vertebra.

Our results were compared, when possible, with existing published experimental measurements of AFJ geometry (Table 1). However, it was not possible to carry out the evaluation of the predictability of all best-fitting equations using experimental data from more datasets. This was due to a lack of availability of measurements in the literature, including values of VBHP and facet parameters in the same dataset. Berry et al. (1987) provided data on the values of VBHP and interfacet height in some vertebral levels but a reference system different to Panjabi et al. (1993) was used. The dataset of Boszczyk (1997) provides accurate experimental measurements of VBHP and the main AFJ parameters, but these were obtained using different landmarks to Panjabi et al. (1993). Ebraheim et al. (1997) provided a large quantity of data on facet parameters but no information about the size of the vertebral bodies. Masharawi et al. (2004, 2005, 2008) performed direct measurement on normal spine of facet and vertebral body parameters but the values of VBHP of these studies were not provided.

All correlation coefficients generated in the current study were considerably better than those obtained by Breglia (2006). He correlated interfacet width and inclinations with VBHP by simple linear regression, finding a poor or even no correlation. Our results showed high correlations for these AFJ parameters after nonlinear regressions. The advantages of using nonlinear regressions to fit vertebral data were described in a previous study (Kunkel et al. 2010). Comparisons of the theoretical predictions with the dataset of Cotterill et al. (1986) indicated that the predictions closely agreed with the experimental data.

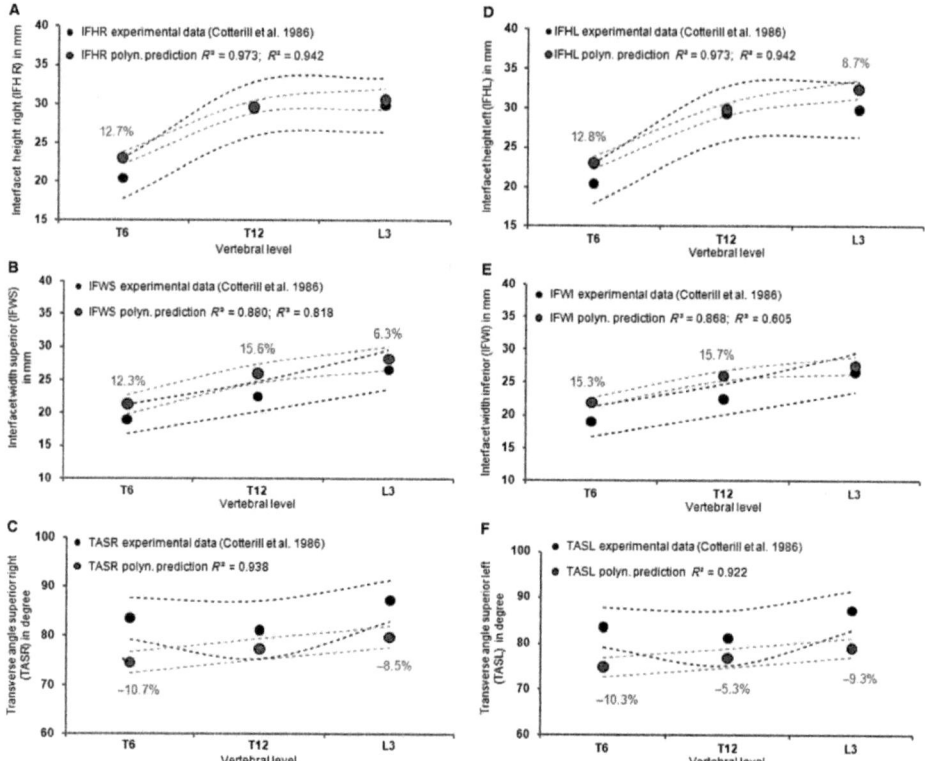

Fig. 5 Comparison of some predicted articular facet joint parameters (IFHR, IFWS, TASR, IFHL, IFWI and TASL) with corresponding experimental data from Cotterill et al. (1986) in selected vertebral levels (T6, T12 and L3). Refer to Fig. 1 for abbreviations. R^2 is provided for thoracic and for lumbar levels. The means and 95% confidence intervals (dotted lines) of both the experimental and the predicted values are shown. Only mean percent errors > 5% for all vertebral levels are shown.

Although the differences in the predictions of lumbar facet parameters were moderate, an excellent correlation between the thoracic facet parameters and VBHP was found (Fig. 5).

It should be emphasized that the predictions using the set of equations generated in the current study are an approximation and should not be extrapolated beyond the limits of the data of Panjabi et al. (1991, 1992). Furthermore, the AFJ show individual and segmental variation (Bernick & Cailliet, 1982; Taylor & Twomey, 1986; Wang & Yang, 2009; Ahmed et al. 1990; Diacinti et al. 1995). Moreover, the facet surfaces that were described here as planar are usually curved (Taylor & Twomey, 1986). Nevertheless, as the two datasets used in this study were provided from in vitro measurements, further investigations are necessary to evaluate the predictability of the regression equations with a dataset from patients.

The advantage of using the generated set of prediction equations (Tables 2 and 3) is the capability to obtain size and orientations of the AFJ considering individual variability from only a single parameter per vertebrae (VBHP) measurable on lateral X-ray. Direct measurement of the main AFJ parameters, considering each vertebral level, cannot be performed in X-rays due to the superposition of several anatomical structures, specifically in the sagittal thoracic region of the spine. Moreover, the lumbar AFJ are difficult to image with X-rays because they are both curved and oblique to the sagittal plane.

Another advantage of using the prediction equations is that they can provide data for parameterized finite element modeling considering patient-specific AFJ morphometry. Thus the geometric congruence between adjacent articular facets could be ensured, and the AFJ changes both between subjects and between vertebral levels maintained. This could avoid the practice adopted in the construction of some models where the AFJ are redirected to force the lower facet of a vertebra to become congruent with the upper facet of the one below (e.g. Maurel et al. 1997; Breglia 2006). Such approximations could produce errors in the modeling of facet contact at different spinal levels.

Conclusion

The present study shows that it is possible to establish useful predictors for human thoracic and lumbar AFJ parameters based on the size of the vertebral body. The generated set of prediction equations enables fast acquisition of geometrical parameters of the AFJ as a function of a single parameter (VBHP), which is measurable in X-rays. As the VBHP is unique for each person and vertebral level, the predicted size and orientations of the AFJ are also specific to an individual. It may be applied for parameterized patient-specific modeling of the spine based on X-ray images alone. Such models make it possible to explore the clinically important mechanical roles of the articular facets in spinal deformities, including scoliosis.

Acknowledgements

This study was financially supported by the German Research Foundation (Wi-1352/12-1).

Conflict of interest statement

The authors of this study did not and will not receive benefits in any form from a commercial party related directly or indirectly to the content of this study.

References

Adams MA, Hutton WC (1983) The mechanical function of the lumbar apophyseal joints. *Spine* 8, 327–330.
Aebi M (2005) The adult scoliosis. *Eur Spine J* 14, 925–948.
Ahmed AM, Duncan NA, Burke DL (1990) The effect of facet geometry on the axial torque-rotation response of lumbar motion segments. *Spine* 15, 391–401.
Aubin C-É, Dansereau J, Parent F, et al. (1997) Morphometric evaluation of personalised 3D reconstructions and geometric models of the human spine. *Med Biol Eng Comput* 35, 611–618.
Bernick S, Cailliet R (1982) Vertebral endplate changes with aging of human vertebrae. *Spine* 7, 97–102.
Berry JL, Moran JM, Berg WS, et al. (1987) A morphometric study of human lumbar and selected thoracic vertebrae. *Spine* 12, 362–367.
Boszczyk BM (1997) *Wirbel und Bewegung – Vergleichende Anatomie der Lendenwirbel – speziell der Wirbelgelenke und Versuch einer Kausalen Analyse*. PhD dissertation. Munchen: University of Munich [in German].
Breglia DP (2006) *Generation of a 3-D Parametric Solid Model of the Human Spine Using Anthropomorphic Parameters*. Masters Dissertation. Athens, OH: Ohio University.
Cotterill PC, Kostuik JP, D'Angelo GD, et al. (1986) An anatomical comparison of the human and bovine thoracolumbar spine. *J Orthop Res* 4, 298–303.
Dai LY (2001) Orientation and tropism of lumbar facet joints in degenerative spondylolisthesis. *Int Orthop* 25, 40–42.
Diacinti D, Acca M, D'Erasmo E, et al. (1995) Aging changes in vertebral morphometry. *Calcif Tissue Int* 57, 426–429.
Ebraheim NA, Xu R, Muhammad A, et al. (1997) The quantitative anatomy of the thoracic facet and the posterior projection of its inferior facet. *Spine* 22, 1811–1817.
Goel VK, Weinstein JN (1990) *Biomechanics of the Spine: Clinical and Surgical Perspective*, pp.14–15. Boca Raton, FL: CRC Press.
Kunkel ME, Schmidt H, Wilke HJ (2010) Prediction equations for human thoracic and lumbar vertebral morphometry. *J Anat* 216, 320–328.
Laporte S, Mitton D, Ismael B, et al. (2000) Quantitative morphometric study of thoracic spine. A preliminary parameters statistical analysis. *Eur J Orthop Surg Traumatol* 10, 85–91.
Lavaste F, Skalli W, Robin S, et al. (1992) Three-dimensional geometrical and mechanical modelling of the lumbar spine. *J Biomech* 25, 1153–1164.
Lorenz M, Patwardhan A, Vanderby R (1983) Load-bearing characteristics of lumbar facets in normal and surgically altered spinal segments. *Spine* 8, 122–130.
Masharawi Y, Rothschild B, Dar G, et al. (2004) Facet orientation in the thoracolumbar spine: three-dimensional anatomic and biomechanical analysis. *Spine* 29, 1755–1763.
Masharawi Y, Rothschild B, Salame K, et al. (2005) Facet tropism and interfacet shape in the thoracolumbar vertebrae. *Spine* 30, E281–E292.
Masharawi Y, Dar G, Peleg S, et al. (2007a) Lumbar facet anatomy changes in spondylolysis: a comparative skeletal study. *Eur Spine J* 16, 993–999.
Masharawi Y, Alperovitch-Najenson D, Steinberg N, et al. (2007b) Lumbar facet orientation in spondylolysis: a skeletal study. *Spine* 32, E176–E180.
Masharawi Y, Salame K, Mirovsky Y, et al. (2008) Vertebral body shape variation in the thoracic and lumbar spine: characterization of its asymmetry and wedging. *Clin Anat* 21, 46–54.
Maurel N, Lavaste F, Skalli W (1997) A three-dimensional parameterized finite element model of the lower cervical spine. Study of the influence of the posterior articular facets. *J Biomech* 30, 921–931.
Onan OA, Hipp JA, Heggeness MH (1998) Use of computed tomography image processing for mapping of human cervical facet surface geometry. *Med Eng Phys* 20, 77–81.
Panjabi MM, Takata K, Goel V, et al. (1991) Thoracic human vertebrae – quantitative three-dimensional anatomy. *Spine* 16, 889–901.
Panjabi MM, Goel V, Oxland T, et al. (1992) Human lumbar vertebrae – quantitative three-dimensional anatomy. *Spine* 17, 299–306.
Panjabi MM, Oxland T, Takata K, et al. (1993) Articular facets of the human spine – quantitative three-dimensional anatomy. *Spine* 18, 1298–1310.

Parent S, Labelle H, Skalli W, et al. (2002) Morphometric analysis of anatomic scoliotic specimens. *Spine* 27, 2305–2311.

Petit Y, Dansereau J, Labelle H, et al. (1998) Estimation of 3D location and orientation of human vertebral facet joints from standing digital radiographs. *Med Biol Eng Comput* 36, 389–394.

Pomero V, Mitton D, Laporte S, et al. (2004) Fast accurate stereographic 3D-reconstruction of the spine using a combined geometric and statistic model. *Clin Biomech* 19, 240–247.

van Schaik JPJ, Verbiest H, Frans DJ (1985) The orientation of laminae and facet joints in the lower lumbar spine. *Spine* 10, 59–63.

Schmidt H, Heuer F, Wilke HJ (2008a) Interaction between finite helical axes and facet joint forces under combined loading. *Spine* 33, 2741–2748.

Schmidt H, Heuer F, Claes L, et al. (2008b) The relation between the instantaneous center of rotation and facet joint forces – A finite element analysis. *Clin Biomech* 23, 270–278.

Schmidt H, Midderhoff S, Adkins K, et al. (2009) The effect of different design concepts in lumbar total disc arthroplasty on the range of motion, facet joint forces and instantaneous center of rotation of a L4-5 segment. *Eur Spine J* 18, 1695–1705.

Scoles PV, Linton AE, Latimer B, et al. (1988) Vertebral body and posterior element morphology: the normal spine in middle life. *Spine* 13, 1082–1086.

Shirazi-Adl A (1991) Finite-element evaluation of contact loads on facets of an L2-L3 lumbar segment in complex loads. *Spine* 16, 533–541.

Shirazi-Adl A (1994) Nonlinear stress analysis of the whole lumbar spine in torsion – mechanics of facet articulation. *J Biomech* 27, 289–299.

Skalli W, Robin S, Lavaste F, et al. (1993) A biomechanical analysis of short segment spinal fixation using a three-dimensional geometric and mechanical model. *Spine* 18, 536–545.

Taylor JR, Twomey LT (1986) Age changes in lumbar zygapophyseal joints: observations on structure and function. *Spine* 11, 739–745.

Wang J, Yang X (2009) Age-related changes in the orientation of lumbar facet joints. *Spine* 34, E596–E598.

White AA, Panjabi MM (1990) *Clinical Biomechanics of the Spine*, pp. 39–40. Philadelphia: Lippincott.

Zander T, Rohlmann A, Klöckner C, et al. (2003) Influence of graded facetectomy and laminectomy on spinal biomechancis. *Eur Spine J* 12, 427–434.

Journal of Anatomy

J. Anat. (2011)   doi: 10.1111/j.1469-7580.2011.01397.x

Morphometric analysis of the relationships between intervertebral disc and vertebral body heights: an anatomical and radiographic study of the human thoracic spine

Maria E. Kunkel,[1] Andrea Herkommer,[1] Michael Reinehr,[2] Tobias M. Böckers[2] and Hans-Joachim Wilke[1]

[1]*Institute of Orthopaedic Research and Biomechanics, University of Ulm, Ulm, Germany*
[2]*Institute of Anatomy and Cell Biology, University of Ulm, Ulm, Germany*

Abstract

The main aim of this study was to provide anatomical data on the heights of the human intervertebral discs for all levels of the thoracic spine by direct and radiographic measurements. Additionally, the heights of the neighboring vertebral bodies were measured, and the prediction of the disc heights based only on the size of the vertebral bodies was investigated. The anterior (ADH), middle (MDH) and posterior heights (PDH) of the discs were measured directly and on radiographs of 72 spine segments from 30 donors (age 57.43 ± 11.27 years). The radiographic measurement error and the reliability of the measurements were calculated. Linear and non-linear regression analyses were employed for investigation of statistical correlations between the heights of the thoracic disc and vertebrae. Radiographic measurements displayed lower repeatability and were shorter than the anatomical ones (approximately 9% for ADH and 37% for PDH). The thickness of the discs varied from 4.5 to 7.2 mm, with the MDH approximately 22.7% greater. The disc heights showed good correlations with the vertebral body heights (R^2, 0.659–0.835, P-values < 0.005; ANOVA), allowing the generation of 10 prediction equations. New data on thoracic disc morphometry were provided in this study. The generated set of regression equations could be used to predict thoracic disc heights from radiographic measurement of the vertebral body height posterior. For the creation of parameterized models of the human thoracic discs, the use of the prediction equations could eliminate the need for direct measurement on intervertebral discs. Moreover, the error produced by radiographic measurements could be reduced at least for the PDH.

Key words: anatomical measurement; disc morphometry; intervertebral disc height; radiographic measurement; thoracic vertebrae.

Introduction

The thoracic spine is the most common site for spinal deformities such as kyphosis and scoliosis (Lord et al. 1995). Despite this, and in contrast to cervical and lumbar discs, the morphometry of the adult thoracic intervertebral disc (TIVD) has received limited attention and until now relatively few data have been available in the current literature. For example, accurate anatomical data on the heights of the TIVD including all levels of the thoracic spine of a representative adult population are very scarce. Previous studies showed limitations either in accuracy, study population, parameters recorded or disc level. Anatomical data on TIVD are a requirement for both the development of new spinal implants and for the creation of mathematical models of the human spine.

Direct measurement on specimens is the best method for extracting morphometric data from anatomical structures. However, relatively few studies on TIVD morphometry have been carried out due to the difficulty in obtaining intact human specimens. Hurxthal (1968) and Manns et al. (1986) measured anterior disc height (ADH) using radiographs of

Correspondence
Hans-Joachim Wilke, Institute of Orthopaedic Research and Biomechanics, Helmholtzstrasse 14, D-89081 Ulm, Germany. T: 0049 731 500 55320; F: 0049 731 500 55302; E: hans-joachim.wilke@uni-ulm.de

Accepted for publication *27 April 2011*

© 2011 The Authors
Journal of Anatomy © 2011 Anatomical Society of Great Britain and Ireland

female patients, but only a limited number of thoracic levels (from T5–6 to T11–12) were investigated. Todd & Pyle (1928) measured only ADH and only male cadavers were used, while the age distribution for this sample was not reported; Pooni et al. (1986) used only a few elderly cadavers between 73 and 85 years old, but the data were presented only as a percentage of the total spine height; radiographic measurements by Goh et al. (1999) and Giles & Singer (2000) were used to investigate thoracic kyphosis, but the ADH and posterior heights (PHD) of the disc were not provided and only a segmental trend was reported. Some of these measurements were performed on plain radiographs considering the superior and inferior vertebral corners (Pooni et al. 1986; Goh et al. 1999; Giles & Singer, 2000). Whereas it has the advantage of eliminating the need for sample preparation, in some reports on lumbar discs the accuracy and repeatability of radiographic measurements of disc heights have been questioned (Pope et al. 1977; Andersson et al. 1981). This was due to a lack of the requisites needed to perform geometric measurements with relative accuracy such as the use of a standard vertebral position, control of the film–specimen–focus distances and optimal visualization of the bony landmarks. Furthermore, in some investigations on TIVD morphometry, errors due to radiographic magnification bias or the inter- and intraobserver reliability of the radiographic measurements were not taken into account (Hurxthal, 1968; Manns et al. 1986; Pooni et al. 1986). The error of radiographic measurement of the heights of TIVDs was never accurately investigated.

Statistical correlations between the main anatomical dimensions of the human vertebral structures have been quantified in previous studies (Scoles et al. 1988; Lavaste et al. 1992; Laporte et al. 2000; Breglia, 2006; van der Houwen et al. 2010). Recently, thoracic and lumbar vertebral morphometry was predicted with reasonable accuracy using only the dimension of the vertebral body heights measured on lateral radiographs and a set of regression equations (Kunkel et al. 2010, 2011). A similar method could be used for the prediction of TIVD dimensions as an alternative to anatomical or radiographic measurements. However, the relationships between morphometric dimensions of the TIVD and vertebrae have never been investigated. To the authors' knowledge, to date no report has investigated the possibility of establishing useful predictors for TIVD dimensions based only on the size of the vertebral bodies. Such a method could provide data for patient-specific modeling of the spine where the shape and size of the TIVD need to be considered.

The main aim of this study was to provide anatomical data on the heights of the human intervertebral discs for all levels of the thoracic spine by direct and radiographic measurements. The radiographic measurements error was also estimated. Additionally, the heights of the neighboring vertebral bodies were measured, and the prediction of the disc heights based only on the size of the vertebral bodies was investigated.

Materials and methods

Study sample and parameters

Seventy-two isolated spine segments (each segment includes a vertebral pair and intervertebral disc between the vertebrae)

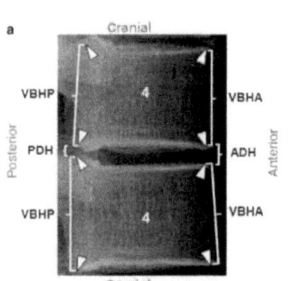
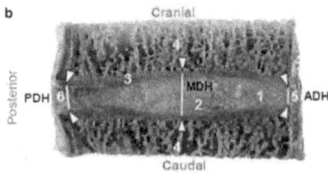

Morphometric parameters

ADH — Anterior disc height
MDH — Middle disc height
PDH — Posterior disc height
VBHA — Vertebral body height anterior
VBHP — Vertebral body height posterior

Morphometric indices

Relative disc height index (I_1)
$I_1 = (ADH + PDH) / (VBHa + VBHp)$
Disc convexity index (I_2)
$I_2 = MDH / (ADH + PDH)$
Disc anterioroposterior wedging index (I_3)
$I_3 = ADH / PDH$

Anatomical structures

1 Annulus fibrosus
2 Nucleus pulposus
3 Cartilaginous end-plate
4 Vertebral body
5 Anterior longitudinal ligament
6 Posterior longitudinal ligament

Fig. 1 Schematic representation of the parameters and indices that were considered for the morphometric analyses. The anatomical landmarks used are indicated by white arrowheads. (a) In the lateral conventional radiographs of the spinal segments all listed parameters, with the exception of the MDH, were measured. (b) In the sagittal sections of the specimens just the three heights of the intervertebral discs were measured. Images are from a thoracic segment (T9–10) of a 57-year-old female donor. ADH, anterior disc height; MDH, middle disc height; PDH, posterior disc height; VBHA, vertebral body height anterior; VBHP, vertebral body height posterior.

© 2011 The Authors
Journal of Anatomy © 2011 Anatomical Society of Great Britain and Ireland

from a total of 30 human spines were examined. For morphometric measurements and statistical analyses six segments were available for each spinal level from C7–T1 to T11–12. Thirty-seven spine segments were from 15 females (mean age of 58.67 ± 10.74 years, range: 43–80 years) and 35 were from 15 males (mean age of 56.20 ± 11.65 years, range: 37–79 years). The discs were not classified into age or gender groups. Because there is not a grading system for the anatomical or radiographic identification of degenerative change in TIVDs, an overview of features of the discs (nucleus and annulus) and vertebrae (end-plate and vertebral body) of the thoracic spine was proposed based on previous studies on lumbar discs; macroscopic classification schemas (Nachemson, 1979; Galante, 1967; Thompson et al. 1990; Adams et al. 1996; Wilke et al. 2006) and studies taking into account pathological changes to lumbar discs (Friberg & Hirsch, 1949; Hirsch & Schajowicz, 1953; Vernon-Robert & Pirie, 1977). A set of five morphometric parameters was measured both on lateral radiographs obtained in controlled conditions and directly on sagittal sections of the specimens (Fig. 1). From these measurements, three adimensional morphometric indices were calculated based on previous methods for lumbar discs (Twomey & Taylor, 1987; Amonoo-Kuofi, 1991). The junctional segment C7-T1 was also included for the observation of anatomical variations in the cervicothoracic discs in the transition from mobile cervical to rigid thoracic spine.

Specimen preparation, radiographic imaging and measurement

The spines were dissected into individual segments free of soft tissues. The ribs were sawn and segments containing the whole disc attached to its upper and lower vertebrae were stored at −28 °C. Lateral and antero-posterior radiographs were taken using a Faxitron automatic X-ray machine (Hewlett Packard, McMinnville, USA). Each segment was placed in a neutral, standard position and a standard film–focus distance of 60 cm, time of 60 s and a tube voltage of 46.5 kV were used. The X-ray beam was centered on the discs. Potential error due to off-center positioning of the spinal segments from the X-ray was examined and a factor for corrections of differences in magnification was calculated. For the radiographic measurements, individual radiographs of the spinal segments were placed on a viewing table and eight anatomical landmarks representing the four corners on the extreme anterior and posterior margins of the end-plates of the vertebrae were marked using Farfan's method (1973) (Fig. 1a). The disc and vertebral heights were measured using an electronic digital caliper (Mitutoyo, Absolute Digimatic, Tokyo, Japan) with an accuracy of ± 0.05 mm. The radiographs were calibrated by means of a scale placed on the rig close to the specimen. Radiographs with overlay from other structures or of deficient film quality were excluded.

Specimen preparation and anatomical measurement

To perform the direct measurements on the discs, frozen spinal segments were sectioned in the horizontal plane through each of the upper and lower vertebral bodies using a high-precision saw (Exacta; PSI Medical, Grünwald, Germany). The posterior elements were removed and segments containing the whole disc attached to a thick portion of the upper and lower vertebral bodies were maintained. Two rods, each 3.0 mm in diameter and 20 mm in length, were fixed into each vertebral body indicating the frontal and median sagittal plane of the TIVD. To produce sagittal sections of the TIVD, each specimen was frozen in an ice block that was individually mounted into a holder with adjustable height for a saw microtome (Leica SP4000; Leica Microsystems, Wetzlar, Germany). The ice blocks were subjected to sagittal sectioning based on the position of the rods. A sliding vernier caliper (Mitutoyo, Absolute Digimatic, Tokyo, Japan) was used for the measurement of TIVD heights, including the cartilaginous end-plates, by using the previously described anatomical landmarks in addition to two mid-vertebral points located on the superior and inferior end-plates for the measurement of the middle disc height (MDH) (Fig. 1b).

Inter- and intra-observer reliability

Each set of radiographic and anatomical measurements was carried out by two observers. Inter-observer errors were examined by repeating the measurement of all parameters in all radiographs and anatomical specimens. Intra-observer errors were examined by one observer making repeated measurements in 10 radiographs from 10 individual specimens at five different spinal levels with some minutes between each repeated measurement. The measurement precision for each parameter was expressed as a coefficient of variation (CV).

Statistical analysis

Linear regression was used to examine the correlation between the radiographic and anatomical measurements of the ADH and PDH, and to calculate the accuracy of the radiographic measurement in relation to the anatomical one. The heights of the discs and the anterior and posterior vertebral bodies (VBHA and VBHP) were individually regressed against the parameters ADH, VBHA and VBHP by a least-square estimation process using a methodology based on Kunkel et al. (2010, 2011). Linear and non-linear regression analyses were employed to find the best functions to fit each of these parameters in a prediction equation. The parameters ADH, VBHA and VBHP were chosen as predictor variables because they could be measured on radiographs with an acceptable accuracy. This was not the case for the MDH and PDH, which were excluded as predictor variables. An ANOVA was performed to define the significance of the prediction equations ($P < 0.05$) that were evaluated using experimental data of Todd & Pyle (1928).

Results

The grading system proposed to classify the thoracic spinal segments consisted of four grades: (i) no degeneration; (ii) mild degeneration; (iii) moderate degeneration; and (iv) strongly degenerated. The description of the grading scale is as follows. Nucleus: (i) elastic, bright, clear delineation from the annulus; (ii) slightly fibrotic, no clear delineation from the annulus; (iii) fibrous, dry, fissured, discolored, bleeding through cavities, loss of the annulus–nucleus boundary; (iv) fibrous, dry, brownish, brittle, partially replaced by scar tissue. Annulus: (i) concentrically arranged fiber ring plates, regular onion-shaped grain, shiny, sinewy;

Table 1 Anatomical and radiographic measurements of the anterior, middle and posterior human TIVD heights.

	Anterior disc height (ADH)						Posterior disc height (PDH)						Middle disc height (MDH)*		Average disc height **
	Anatomical measurement		Radiographic measurement		Anatomical- radiographic ratio		Anatomical measurement		Radiographic measurement		Anatomical- Radiographic ratio		Anatomical measurement		
Disc level	Mean	SD	Mean	SD	Difference	%	Mean	SD	Mean	SD	Difference	%	Mean	SD	
C7-T1	4.5	0.79	3.96	0.22	0.54	-11.90	4.5	0.51	3.20	0.55	1.30	-28.91	5.6	0.63	4.5
T1-2	4.5	0.77	3.69	0.62	0.81	-18.01	4.3	0.62	3.04	0.56	1.26	-29.40	5.7	0.66	4.4
T2-3	3.4	0.97	3.23	0.40	0.17	-4.89	3.5	0.99	1.95	0.28	1.55	-44.22	5.8	0.79	3.5
T3-4	3.3	0.30	3.07	0.44	0.23	-6.96	3.2	0.49	2.09	0.46	1.11	-34.69	4.4	0.8	3.3
T4-5	3.0	0.74	2.85	0.36	0.15	-5.11	3.3	0.45	2.00	0.21	1.30	-39.54	5.0	0.61	3.2
T5-6	3.5	0.37	3.36	0.29	0.14	-3.93	3.6	0.47	2.10	0.30	1.50	-41.71	4.7	0.91	3.5
T6-7	4.1	0.35	3.70	0.41	0.40	-9.80	4.1	0.65	2.18	0.50	1.92	-46.72	5.5	0.45	4.1
T7-8	4.2	0.97	4.06	0.67	0.14	-3.35	3.6	0.90	2.47	0.50	1.13	-31.44	5.3	0.58	3.9
T8-9	5.6	1.17	4.52	0.43	1.08	-19.22	5.0	0.91	2.49	0.60	2.51	-50.12	6.1	0.89	5.3
T9-10	5.4	1.74	4.88	1.03	0.52	-9.70	4.2	1.10	2.97	0.60	1.23	-29.36	5.5	0.90	4.8
T10-11	7.2	1.21	5.91	0.77	1.29	-17.94	5.8	1.00	3.66	0.70	2.14	-36.83	6.7	1.18	6.5
T11-12	6.0	1.14	5.90	0.70	0.10	-1.67	4.8	1.26	3.23	0.48	1.57	-32.63	6.5	1.15	5.4

Refer to Fig. 1 for abbreviations.
All measurements are in mm.
*No radiographic measurement was performed for the MDH.
**Average disc height is the average of the anatomical measurement of the ADH and PDH at each vertebral level. o = digital.

(ii) sharply contoured, concentrically arranged fiber ring plates (drier appearance); (iii) disordered fibrous structure of the lamellae, fiber faults, fiber fabric ring pronounced dry-looking, sprouting of blood vessels; (iv) ruptures in the annulus, fiber breaks, cracks, fissures, defects. End-plate: (i) well-built hyaline end-plate, even thickness; (ii) hyaline with irregular thickness; (iii) local cartilage defects; (iv) complete destruction of cartilaginous end-plate. Vertebral body: (i) rounded margins; (ii) small projections from the margins; (iii) first osteophytes on the edge < 2 mm; (iv) osteophytes on the anterior vertices > 2 mm. The anatomical and radiographic inspection of the spinal segments selected for this study showed mild to moderate degenerative changes (grades ii and iii).

Inter- and intra-observer reliability of the measurements

Inter-observer reliability showed that measurements of the TIVD heights were better repeated when obtained directly

Table 2 Anterior and posterior human thoracic vertebral body heights.

Vertebral level	Vertebral body height anterior (VBHA)		Vertebral body height posterior (VBHP)		Average vertebral body height*
	Mean	SD	Mean	SD	
C7	13.89	1.42	14.38	2.22	14.14
T1	14.49	1.23	15.28	1.14	14.88
T2	15.01	1.51	16.11	1.67	15.56
T3	15.65	1.85	17.41	1.12	16.53
T4	15.42	1.46	18.15	1.54	16.79
T5	15.84	1.07	17.33	2.16	16.59
T6	16.04	1.43	18.22	1.38	17.13
T7	15.94	1.61	18.67	1.64	17.31
T8	16.99	1.70	20.05	1.77	18.52
T9	18.26	2.12	20.25	2.44	19.26
T10	18.98	1.40	20.35	1.89	19.67
T11	19.60	1.92	22.67	1.38	21.14
T12	20.80	1.96	23.12	1.94	21.96

Refer to Fig. 1 for abbreviations.
All measurements are in mm.
*Average vertebral body height is the average of the VBHA and VBHP at each vertebral level.

Table 3 Morphometric indices derived from measurements on the intervertebral discs and vertebral bodies heights.

Disc level	Relative disc height index (I_1)	Disc : vertebral body height ratio	Disc convexity index (I_2)	% MDH to average disc height	Disc antero-posterior wedging index (I_3)
C7–T1	0.32	1:3.1	0.63	20.55	1.00
T1–2	0.30	1:3.4	0.65	23.11	1.06
T2–3	0.22	1:4.5	0.83	39.65	0.97
T3–4	0.20	1:5.1	0.67	25.17	1.03
T4–5	0.19	1:5.3	0.80	37.50	0.89
T5–6	0.21	1:4.7	0.67	25.58	0.97
T6–7	0.24	1:4.2	0.68	25.95	1.01
T7–8	0.23	1:4.4	0.68	26.75	1.16
T8–9	0.29	1:3.5	0.57	12.56	1.13
T9–10	0.25	1:4.0	0.58	13.74	1.29
T10–11	0.33	1:3.0	0.52	3.50	1.23
T11–12	0.26	1:3.9	0.60	16.77	1.25

Refer to Fig. 1 for abbreviations.
I_1 = (ADH + PDH)/(VBHA + VBHP) based on Amonoo-Kuofi (1991).
I_2 = MDH/(ADH + PDH) based on Twomey & Taylor (1987).
I_3 = ADH/PDH.
The ratio of disc: body was calculated by dividing 1 by the value of I_1.

© 2011 The Authors
Journal of Anatomy © 2011 Anatomical Society of Great Britain and Ireland

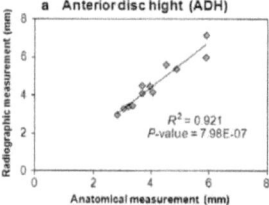
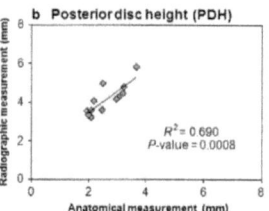

Fig. 2 Correlations between anatomical and radiographic measurements of the anterior (a) and posterior (b) disc heights considering mean values from each disc level from C7–T1 to T11–12. See Table 1 for data summaries. ADH, anterior disc height; PDH, posterior disc height.

from the specimens (CV = 0.79–0.93) than from the radiographs (CV = 0.49–0.82). A lower repeatability was found for radiographic measurement of the PDH (CV = 0.49). For radiographic measurements of the VBHA and VBHP a high repeatability was found (CV = 0.95–0.98). Intra-observer reliability showed that anatomical measurements were reproduced with errors ranging from 1.7 to 6.1% for ADH, 17 to 26.1% for PDH and 1.7 to 5.1% for VBHA and VBHP. Reproducibility of the measurements from repeat radiographs was generally 15% lower.

Intervertebral discs and vertebral body heights

A wide variation in TIVD heights was found in the anatomical measurements (Table 1). Direct measurement of the ADH varied from approximately 4.5 mm at C7–T1, with a gradual decrease towards T4–5 (approximately 3 mm), increasing again caudally to approximately 7.2 mm at T10–11, and decreasing again to approximately 6 mm at T11–12 (Table 1). PDH measured directly on the specimens followed a trend similar to ADH, but from the disc level T7–8 there was an increase of approximately 21% in the values. The average disc height that corresponded to the average of the ADH and PDH at each vertebral level varied from 3.2 to 6.5 mm (mean value of 4.3 ± 1 mm; Table 1). The average vertebral body height varied from 14.2 to 21.96 mm (Table 2). The MDH was on average 22.7% higher than the average disc height (Tables 1 and 3). Radiographic ADH and PDH values were shorter than the anatomical ones (approximately 9% for ADH and 37% for PDH). From these comparisons, a height linear correlation was found for ADH ($R^2 = 0.921$) (Fig. 2a), but only a moderate correlation for PDH ($R^2 = 0.690$) (Fig. 2b).

Morphometric indices

The index I_1 enabled a comparison of the TIVD height with the heights of neighboring vertebral bodies (Table 3). There was a constant relationship between the disc thickness and the vertebral bodies' heights at all levels (ratio disc : body of approximately 1 : 4.1) (Fig. 3). The index I_2 indicated that the ovality of the disc did not follow a trend as one descends the thoracic spine. It was more pronounced at the T2–3 level where MDH was almost 40% of the average disc height (Table 3; Fig. 3). The index I_3 showed that with the exception of the cervicothoracic discs (C7–T1) all discs were wedge-shaped. There was a trend for a posterior wedge configuration from the T7–8 level onwards, the ADH being approximately 21.25% greater than the PDH. However, in the upper and middle thoracic region (from T1–2 to T6–7) both anteriorly and posteriorly minimal wedge shapes were found with a maximum difference between these values of approximately 2.77%. From these morphometric indices,

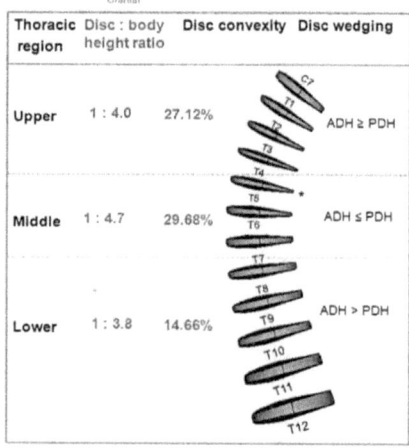

Thoracic region	Disc : body height ratio	Disc convexity	Disc wedging
Upper	1 : 4.0	27.12%	ADH ≥ PDH
Middle	1 : 4.7	29.68%	ADH ≤ PDH
Lower	1 : 3.8	14.66%	ADH > PDH

Fig. 3 Geometric model of the human thoracic discs from C7–T1 to T11–12 disc level, constructed with parameters derived from the morphometric analyses performed in this study. The thinnest disc was found in the disc level T4–5*. The shape of the human TIVDs was determined by the relationships between the anterior, middle and posterior disc heights (ADH, MDH, PDH).

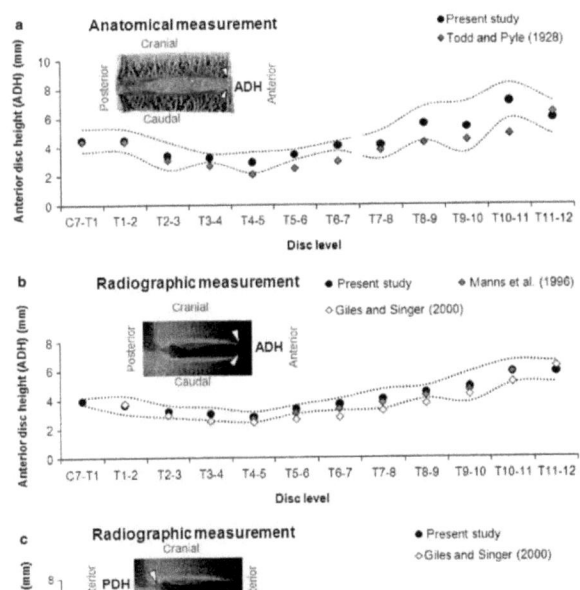

Fig. 4 Comparison of the anterior and posterior heights of the human intervertebral discs obtained in the present study with published data. Mean values of anatomical (ADH; a) and radiographic (ADH and PDH; b,c) measurements. Dotted curves indicate SD of the experimental values of the present study. ADH, anterior disc height; PDH, posterior disc height.

the TIVD could be classified into three distinct regions as an approximation of the vertebral regions of the thoracic spine (Panjabi et al. 1991a; Fig. 3). The upper region of transition from cervical to thoracic from C7–T1 to T3–4 was characterized by a gradual decrease caudally of the ADH and PDH until the thinnest disc at T4–5, with a sliding posterior wending. The middle region from T4–5 to T6–7 showed sliding anterior wedge-shaped discs, and the lower region from the apex of the thoracic spine T7–8 to T11–12 contained more posterior wedge-shaped discs. Comparisons with the literature were provided for measurements of the disc heights (Fig. 4) and vertebral bodies heights (Fig. 5).

Disc and vertebral body height correlations

In general, the heights of the TIVDs presented had good correlations with the vertebral heights, that were significant (R^2 = 0.659–0.835, $P < 0.005$; ANOVA) (Table 4). An exception was the MDH, for which no significant correlations with the vertebral body heights were found ($R^2 < 0.6$, $P > 0.05$). A set of 10 polynomial equations was generated for the prediction of TIVD heights from parameters that could be accurately measured on the radiographs (ADH, VBHA and VBHP; Table 4). The polynomial predictions, using the generated set of regression equations, were generally within or close to the region of the 95% confidence intervals of the experimental data measured in the current study (Fig. 6). The evaluation of the predictability of the regression equations using VBHA and VBHP of the data set of the radiographic measurements of the current study showed that good results could be found (Figs 7 and 8). Using the data set of Todd & Pyle (1928), a comparison of predicted PDH from radiographic ADH showed a greatest error of approximately −13% in the upper and −17% in the

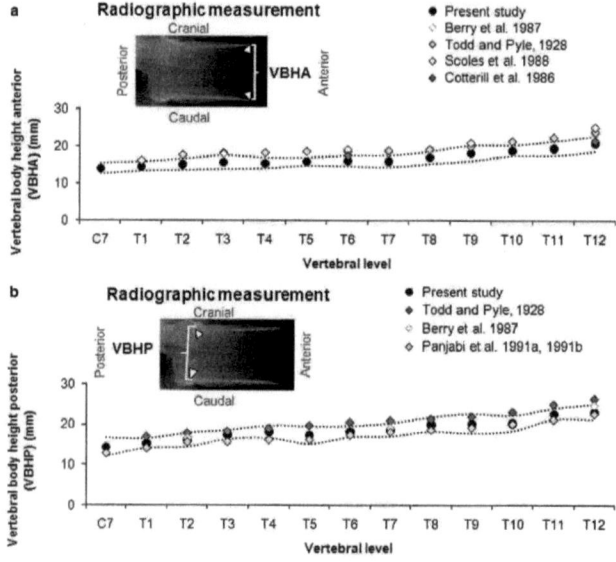

Fig. 5 Comparison of the anterior (a) and posterior (b) heights of the human vertebral bodies obtained in the present study with published data. Means values of radiographic measurements. Dotted curves indicate SD of the experimental values of the present study. VBHA, vertebral body height anterior; VBHP, vertebral body height posterior.

Table 4 Polynomial coefficients (C_1, C_2, C_3 and C_4) for prediction equations of the parameters related to the heights of the TIVD and vertebral bodies for all spinal levels from C7–T1 to T11–12. The vertebral parameters come from radiographic measurements, whereas the disc parameters are from anatomical measurements performed in the current study.

Predictor parameter	Predicted parameter	R^2	P-value	SE	C_1	C_2	C_3	C_4
ADH	MDH	0.726	0.0004	0.37	3.500	0.453	–	–
	PDH	0.906	0.0002	0.27	–2.149	3.067	–0.538	0.037
VBHA	ADH	0.835	0.0017	0.60	544.560	–97.473	5.795	–0.113
	MDH	0.576	0.0640	–	–	–	–	–
	PDH	0.667	0.0260	0.52	360.580	–63.935	3.783	–0.074
	VBHP	0.953	2.7E-06	0.66	–180.00	31.073	–1.646	0.030
VBHP	ADH	0.780	0.0053	0.66	298.43	–46.993	2.561	–0.045
	MDH	0.571	0.0676	–	–	–	–	–
	PDH	0.659	0.0282	0.53	201.190	–32.029	1.705	–0.030
	VBHA	0.943	6.5E-06	0.58	44.199	–5.305	0.286	–0.004

Refer to Fig. 1 for abbreviations.
SE in mm.
The basic form of the prediction equations is $y = C_1 + C_2x + C_3x^2 + C_4x^3$ where y is the value of the parameter to be predicted and x is the value of the predictor parameter on each spinal level.

lower regions of the thoracic spine (Fig. 9a). The predictions of the VBHA and VBHP exhibited a mean percent error < 17% (Fig. 9b,c).

Discussion

In the present study, the heights of human TIVDs from the C7–T1 to T11–12 spinal level were measured directly on the sagittal section of 72 specimens and on their radiographs. The main aim was to provide these anatomical data and estimate the error of the radiographic measurements. Additionally, heights of the neighboring vertebral bodies were measured for the investigation of predictions of TIVD dimensions based on the size of the vertebral bodies.

Disc height is an important dimension often used as a diagnostic tool in orthopedics as well as in mathematical modeling of the human spine. Although Oliver & Middleditch (1991) reported that TIVDs have a nearly uniform thickness, our anatomical observations indicated that there

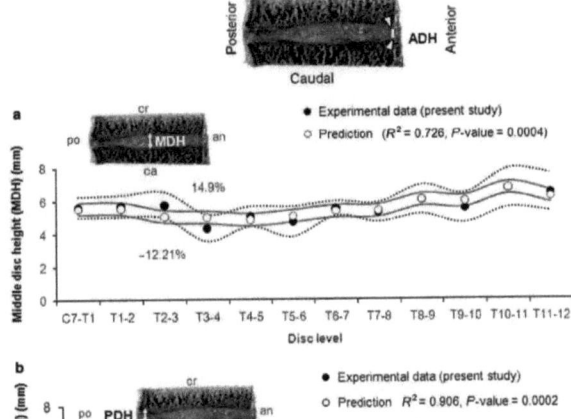

Fig. 6 Anatomical values of the ADH were used for predictions of the parameters MDH (a) and PDH (b) at all levels of the thoracic spine. The predicted values were superimposed on experimental data that were measured in the present study. Dotted and continuous curves indicate SD of the experimental and predicted values, respectively. Mean percent errors of the predictions larger than 10% are indicated. ADH, anterior disc height; an, anterior; ca, caudal; cr, cranial; MDH, middle disc height; PDH, posterior disc height; po, posterior.

is no single exact disc height because the planes that bound the TIVD superiorly and inferiorly were not parallel (Fig. 1). A geometric model of the thoracic discs based on these parameters provides a better visualization of this variation, which occurs in different parts of the same disc as well as in different regions of the thoracic spine (Fig. 3). However, this may be due to the age of the subjects (Vernon-Robert & Pirie, 1977; Twomey & Taylor, 1987) or to other factors such as loss of disc height in cases of scoliosis or disc herniation. Possible sources of error in these measurements could also be due to *post mortem* changes or the degree of disc degeneration (Peacock, 1952; Walmsley, 1953; White & Panjabi, 1990; Goh et al. 2000).

In the current study, the two main sources of ambiguity found in radiographic measurements of disc heights (the disc orientation with respect to the central X-ray beam, and the estimation of differences among different observers) were minimized using the recommendations of Pope et al. (1977) and Andersson et al. (1981). The difficulty in identifying the bony landmarks was overcome by strictly controlling the vertebral position, preserving the relationships between the TIVD and the vertebral bodies. However, even when the specimens were radiographed under these controlled and standard conditions the same degree of accuracy that was seen for the anatomical measurement was still not achieved (Table 1; Fig. 2). This was due to the difficulty in the identification of the vertebral bony landmarks, particularly where many overlapping shadows were found, for example in the measurements of the PDH in the upper thoracic spine. It was not possible to compare radiographic and anatomical measurement of the MDH because the midline distance between the oval radiographic images of the proximal and distal vertebral end-plates could not be identified (Edmondston et al. 1999). Our radiographic ADH and PDH values were shorter than the anatomical ones, probably because the anatomical measurements included the cartilaginous end-plates that cannot be readily identified on radiographs.

Comparison of our direct and radiographic measurements with other studies on TIVD was difficult due to the fact that there are few comparative data in the literature (e.g. no published data related to MDH were found). For ADH, a good agreement was found with anatomical values of Todd & Pyle (1928; Fig. 4a) and radiographic values of Manns et al. (1986) and Giles & Singer (2000) (Fig. 4b); although the same was not found for the radiographic PDH values compared with Giles & Singer (2000) (Fig. 4c). The small variations between the radiographic measurements could

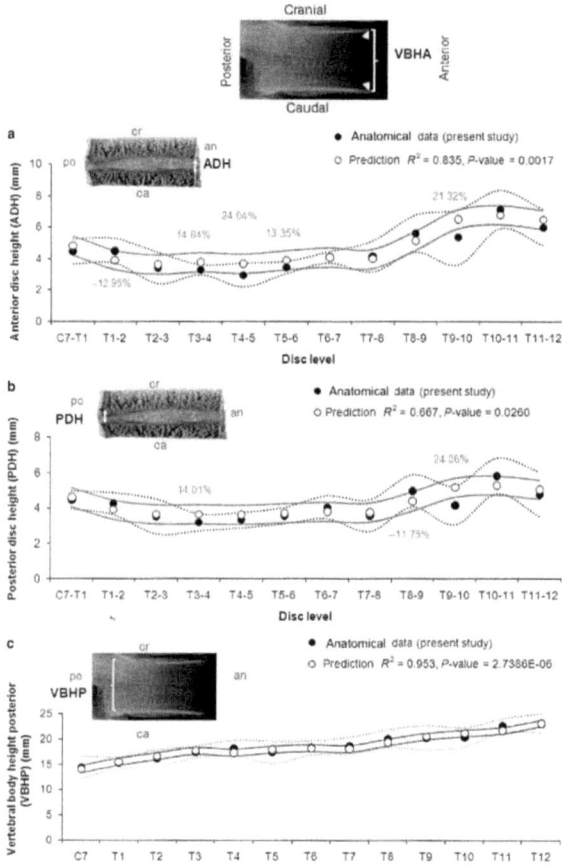

Fig. 7 Radiographic values of the VBHA were used for predictions of the parameters ADH (a), PDH (b) and VBHP (c) at all levels of the thoracic spine. The predicted values were superimposed on anatomical data that were measured in the present study. Dotted and continuous curves indicate SD of the experimental and predicted values, respectively. Mean percent errors of the predictions larger than 10% are indicated. ADH, anterior disc height; an, anterior; ca, caudal; cr, cranial; PDH, posterior disc height; po, posterior; VBHA, vertebral body height anterior; VBHP, vertebral body height posterior.

be due to radiographic magnification bias, positioning errors and distortion due to parallax effects.

As expected, the radiographic measurements of the thoracic vertebral bodies heights showed very good agreement with other studies where these values have already been well established (Todd & Pyle, 1928; Cotterill et al. 1986; Berry et al. 1987; Scoles et al. 1988; Panjabi et al. 1991a,b) (Fig. 5). However, the cartilaginous endplates were not considered for this measurement, and this could slightly interfere in this measurement. Although this was not a major goal of our study, accurate measurements of these parameters were necessary so that we could perform analysis of the correlations with the measurements of disc height. The morphometric indices for TIVD showed a ratio of disc to vertebral body height of approximately 1 : 4.1, with a progressive increase in the spine motion in the sagittal plane in the lower region (1 : 3.8) where the disc height was greater and the thoracic segments were less impeded by the constraint of the thoracic cage. This agrees with Kapandji (1985) and White & Panjabi (1990) who reported the thoracic region as the least mobile of the spine.

Due to the fact that TIVD heights provided in this study were obtained from *in vitro* measurements of isolated spine

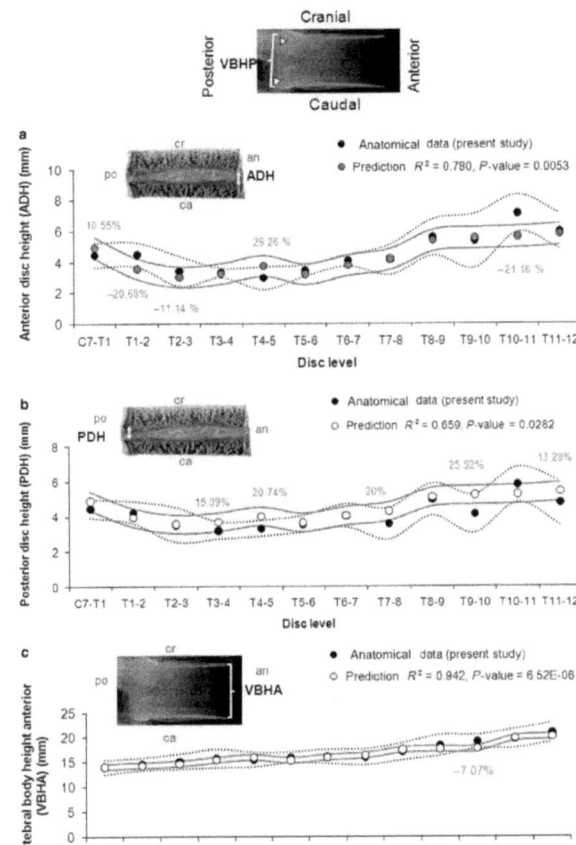

Fig. 8 Radiographic values of the VBHP were used for polynomial predictions of the parameters ADH (a), PDH (b) and VBHA (c) at all levels of the thoracic spine. The predicted values were superimposed on anatomical data that were measured in the present study. Dotted and continuous curves indicate SD of the experimental and predicted values, respectively. Mean percent errors of the predictions larger than 10% are indicated. ADH, anterior disc height; an, anterior; ca, caudal; cr, cranial; PDH, posterior disc height; po, posterior; VBHA, vertebral body height anterior; VBHP, vertebral body height posterior.

segments of cadavers, the axial load applied to the thoracic spine due to bodyweight could not be considered. Therefore, the measured values in the current study should be slightly larger than the radiographic values of a living person in a sitting or standing position. This could be an important point from both arthroplasty and modeling standpoints.

Using the set of prediction equations generated in this study it was possible to estimate heights of the thoracic discs from initial radiographic measurement of the vertebral heights (Figs 6–8). ADH could be predicted, with a largest error of approximately 26%, from measurements of the VBHA or VBHP. MDH could only be measured with statistical significance from the ADH measurements (largest error of approximately 15%). For estimation of PDH, both ADH and vertebral heights provided good predictions. From the measurement of the vertebral height were predicted values of PDH with approximately 26% error, which was less than the radiographic measurement of this parameter in all thoracic levels (Table 1; Figs 7 and 8). The values of VBHA and VBHP allowed very good predictions with errors of less than 10%. The generated equations are thus valid under these limited circumstances. It is also important to point out that the errors caused by the measurement of radiographic vertebral bodies may lead to errors in predicting the height of the disc. Moreover, further investigations

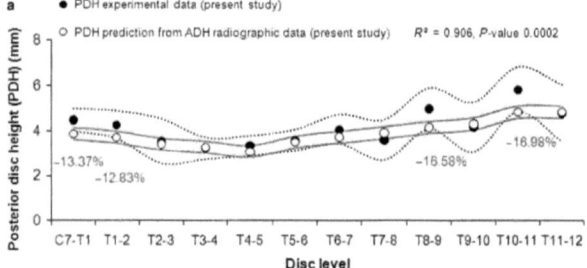

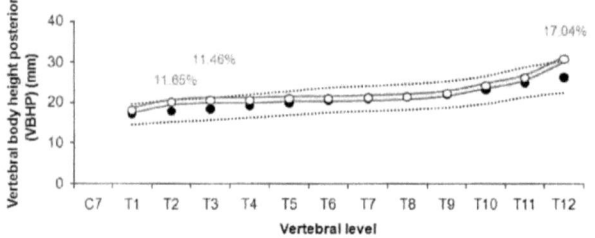

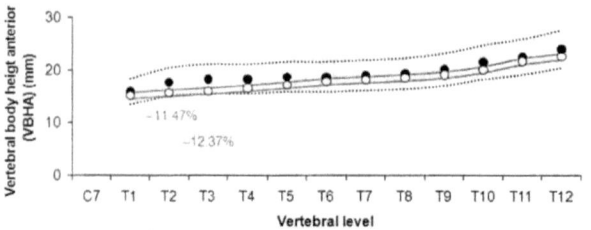

Fig. 9 Comparison of some predicted parameters (PDH, VBHP and VBHA) from anatomical measurements of Todd & Pyle (1928) with corresponding radiographic measurements of the current study. The predicted values were superimposed on experimental data. Dotted and continuous curves indicate SD of the experimental and predicted values, respectively. Mean percent errors of the predictions larger than 10% are indicated. ADH, anterior disc height; PDH, posterior disc height; VBHA, vertebral body height anterior; VBHP, vertebral body height posterior.

are necessary to evaluate the predictability of the regression equations with a data set from patients.

This study provided an accurate and comprehensive data base to describe the geometry of the TIVD. This may serve as an anthropometric reference for mathematical modeling, as well as for anatomical and biomechanical studies of the human spine, where dimensions and relations of spinal bony segments in the mid-sagittal plane are of importance.

Conclusion

The current increased interest in spinal implants and biomechanical models of spinal deformation, including scoliosis, calls for a detailed knowledge of its anatomy and of relationships between the disc and the vertebrae. The present study contributes by providing new anatomical data on thoracic disc morphometry, particularly the disc heights and their relationships with thoracic vertebral body heights. These data are important for understanding the biomechanics and morphology of the spine. The generated set of prediction equations quantitatively describe these relationships and could be used to produce morphometric data on thoracic disc. For the creation of parameterized models of the human thoracic discs, the use of the prediction equations could eliminate the need for direct measurement on intervertebral discs. Moreover, the error produced by radiographic measurements could be reduced at least for the PDH.

Acknowledgement

This study was financially supported by the German Research Foundation (Wi-1352/12-1).

Conflict of interest statement

Each author of this study did not and will not receive benefits in any form from a commercial party related directly or indirectly to the content of this study.

References

Adams MA, McNally DS, Dolan P (1996) Stress distributions inside intervertebral discs. The effects of age and degeneration. *J Bone Joint Surg Br* **78**, 965–972.

Amonoo-Kuofi HS (1991) Morphometric changes in the heights and anteroposterior diameters of the lumbar intervertebral disc with age. *J Anat* **175**, 159–168.

Andersson GBJ, Schultz A, Nathan A, et al. (1981) Roentgenographic measurement of lumbar intervertebral disc height. *Spine* **6**, 154–157.

Berry JL, Moran JM, Berg WS, et al. (1987) A morphometric study of human lumbar and selected thoracic vertebrae. *Spine* **12**, 362–367.

Breglia DP (2006) *Generation of a 3-D parametric solid model of the human spine using anthropomorphic parameters*. Master dissertation. Ohio: Ohio University.

Cotterill PC, Kostuik JP, D'Angelo GD, et al. (1986) An anatomical comparison of the human and bovine thoracolumbar spine. *J Orthop Res* **4**, 298–303.

Edmondston SJ, Price RI, Valente B, et al. (1999) Measurement of vertebral body heights: ex vivo comparisons between morphometric X-ray absorptiometry, morphometric radiography and direct measurements. *Osteoporos Int* **10**, 7–13.

Farfan HF (1973) *Mechanical disorders of the low back*. Philadelphia: Lea and Febiger.

Friberg S, Hirsch C (1949) Anatomical and clinical studies on lumbar disc degeneration. *Acta Orthop Scand* **19**, 222–242.

Galante JO (1967) Tensile properties of the human lumbar annulus fibrosus. *Acta Orthop Scand (Suppl)* **100**, 1–91.

Giles LGF, Singer KP (2000) *Clinical anatomy and management of thoracic spine pain*, pp. 25–27. Oxford: Butterworth-Heinemann.

Goh S, Price RI, Leedman PJ, et al. (1999) The relative influence of vertebral body and intervertebral disc shape on thoracic kyphosis. *Clin Biomech* **14**, 439–448.

Goh S, Tan C, Price SRI, et al. (2000) Influence of age and gender on thoracic vertebral body shape and disc degeneration: an MR investigation of 169 cases. *J Anat* **197**, 647–657.

Hirsch MD, Schajowicz MD (1953) Studies on structural changes in the lumbar annulus fibrosus. *Acta Orthop Scand* **22**, 184–231.

van der Houwen EB, Baron PH, Veldhuizen AG, et al. (2010) Geometry of the intervertebral volume and vertebral endplates of the human spine. *Ann Biomed Eng* **38**, 33–40.

Hurxthal LM (1968) Measurement of anterior vertebral compressions and biconcave vertebrae. *Am J Roentgenol Rad Ther Nucl Med* **103**, 635–644.

Kapandji IA (1985) *Rumpf und Wirbelsäule*. Stuttgart: Ferdinand Enke.

Kunkel ME, Schmidt H, Wilke HJ (2010) Prediction equations for human thoracic and lumbar vertebral morphometry. *J Anat* **216**, 320–328.

Kunkel ME, Schmidt H, Wilke HJ (2011) Prediction of the human thoracic and lumbar articular facet joint morphometry from radiographic images. *J Anat* **218**, 191–201.

Laporte S, Mitton D, Ismael B, et al. (2000) Quantitative morphometric study of thoracic spine. A preliminary parameters statistical analysis. *Eur J Orthop Surg Traumatol* **10**, 85–91.

Lavaste F, Skalli W, Robin S, et al. (1992) Three-dimensional geometrical and mechanical modelling of the lumbar spine. *J Biomech* **25**, 1153–1164.

Lord MJ, Ogden JA, Ganey TM (1995) Postnatal development of the thoracic spine. *Spine* **20**, 1692–1698.

Manns RA, Haddaway MJ, McCall IW, et al. (1986) The relative contribution of disc and vertebral morphometry to the angle of kyphosis in asymptomatic subjects. *Clin Radiol* **51**, 258–262.

Nachemson AL, Schultz AB, Berkson MH (1979) Mechanical properties of human lumbar spine motion segments. Influence of age, sex, disc level and degeneration. *Spine* **4**, 1–8.

Oliver J, Middleditch A (1991) *Functional anatomy of the spine*, pp. 23–24. Oxford: Butterworth-Heinenmann.

Panjabi MM, Takata K, Goel V, et al. (1991a) Thoracic human vertebrae – quantitative three-dimensional anatomy. *Spine* **16**, 888–901.

Panjabi MM, Duranceau J, Goel V, et al. (1991b) Cervical human vertebrae quantitative three-dimensional anatomy of the middle and lower regions. *Spine* **16**, 861–869.

Peacock A (1952) Observations on the postnatal structure of the intervertebral disc in man. *J Anat* **86**, 162–179.

Pooni JS, Hukins DWL, Harris PF, et al. (1986) Comparison of the structure of human intervertebral discs in the cervical, thoracic and lumbar regions of the spine. *Surg Radiol Anat* **8**, 175–182.

Pope MH, Hanley EN, Matteri RE, et al. (1977) Measurement of intervertebral disc space height. *Spine* **2**, 282–286.

Scoles PV, Linton AE, Buce L, et al. (1988) Vertebral body and posterior element morphology: the normal spine in middle life. *Spine* **13**, 1082–1086.

Thompson JP, Pearce RH, Schechter MT, et al. (1990) Preliminary evaluation of a scheme for grading the gross morphology of the human intervertebral disc. *Spine* **15**, 411–415.

Todd TW, Pyle SI (1928) A quantitative study of the vertebral column by direct and roentgenoscopic methods. *Am J Phys Anthropol* **XII**, 321–338.

Twomey L, Taylor JR (1987) Degenerative age changes in lumbar vertebrae and intervertebral discs. *Clin Orthop* **224**, 97–104.

Vernon-Robert B, Pirie CJ (1977) Degenerative changes in the intervertebral discs of the lumbar spine and their sequelae. *Rheumatol Rehab* **16**, 13–21.

Walmsley R (1953) The development and growth of the intervertebral disc. *Edinb Med J* **60**, 341–364.

White AA, Panjabi MM (1990) *Clinical Biomechanics of the Spine*. Philadelphia: J. B. Lippincott.

Wilke HJ, Rohlmann F, Neidlinger-Wilke C, et al. (2006) Validity and interobserver agreement of a new radiographic grading system for intervertebral disc degeneration: part I. Lumbar spine. *Eur Spine J* **15**, 720–730.

i want morebooks!

Buy your books fast and straightforward online - at one of world's fastest growing online book stores! Environmentally sound due to Print-on-Demand technologies.

Buy your books online at
www.get-morebooks.com

Kaufen Sie Ihre Bücher schnell und unkompliziert online – auf einer der am schnellsten wachsenden Buchhandelsplattformen weltweit! Dank Print-On-Demand umwelt- und ressourcenschonend produziert.

Bücher schneller online kaufen
www.morebooks.de

VDM Verlagsservicegesellschaft mbH
Heinrich-Böcking-Str. 6-8
D - 66121 Saarbrücken

Telefon: +49 681 3720 174
Telefax: +49 681 3720 1749

info@vdm-vsg.de
www.vdm-vsg.de

Printed by Books on Demand GmbH, Norderstedt / Germany